Graphing Calculator Manual

Judith A. Penna
Indiana University Purdue University Indianapolis

Elementary Algebra Graphs and Models

Marvin L. Bittinger
Indiana University Purdue University Indianapolis

David J. Ellenbogen
Community College of Vermont

Barbara L. Johnson
Indiana University Purdue University Indianapolis

Boston San Francisco New York
London Toronto Sydney Tokyo Singapore Madrid
Mexico City Munich Paris Cape Town Hong Kong Montreal

Reproduced by Pearson Addison-Wesley from electronic files supplied by the author.

Publishing as Pearson Addison-Wesley, 75 Arlington Street, Boston, MA 02116

ISBN 0-321-19381-4

1 2 3 4 5 6 QEP 08 07 06 05 04

Contents

The TI-83, TI-83 Plus, and TI-84 Plus Graphics Calculators

Chapter 1
Introduction to Algebraic Expressions

GETTING STARTED

Press ON to turn on the TI-83, TI-83 Plus, or TI-84 Plus graphing calculator. (ON is the key at the bottom left-hand corner of the keypad.) You should see a blinking rectangle, or cursor, on the screen. If you do not see the cursor, try adjusting the display contrast. To do this, first press 2nd. (The 2nd key is in the left column of the keypad. It is yellow on the TI-83 and the TI-83 Plus and blue on the TI-84 Plus.) Then press and hold △ to increase the contrast or ▽ to decrease the contrast.

To turn the calculator off, press 2nd OFF. (OFF is the second operation associated with the ON key.) The calculator will turn itself off automatically after about five minutes without any activity.

Press MODE to display the MODE settings. Initially you should select the settings on the left side of the display. The screen is shown here as it appears on the TI-83 and TI-83 Plus. The mode settings on the TI-84 Plus will appear in a different font and a clock will also appear at the bottom of the screen.

To change a setting on the Mode screen use ▽ or △ to move the blinking cursor to the line of that setting. Then use ▷ or ◁ to move the cursor to the desired setting and press ENTER. Press CLEAR or 2nd QUIT to leave the MODE screen. (QUIT is the second operation associated with the MODE key.) Pressing CLEAR or 2nd QUIT will take you to the home screen where computations are performed.

The TI-83, TI-83 Plus, and TI-84 Plus graphing calculators are very similar in many respects. For that reason, most of the keystrokes and instructions presented in this section of the graphing calculator manual will apply to all three calculators. Where they differ, keystrokes and instructions for using the TI-83 will be given first, followed by those for the TI-83 Plus and the TI-84 Plus.

It will be helpful to read the Introduction to the Graphing Calculator on pages 4 and 5 of your textbook as well as the Getting Started section of your graphing calculator Guidebook before proceeding.

EVALUATING EXPRESSIONS

To evaluate expressions we substitute values for the variables.

Section 1.1, Example 4 Use a graphing calculator to evaluate $3xy + x$ for $x = 65$ and $y = 92$.

Enter the expression in the calculator, replacing x with 65 and y with 92. Press 3 × 6 5 × 9 2 + 6 5 ENTER. The calculator returns the value of the expression, 18,005.

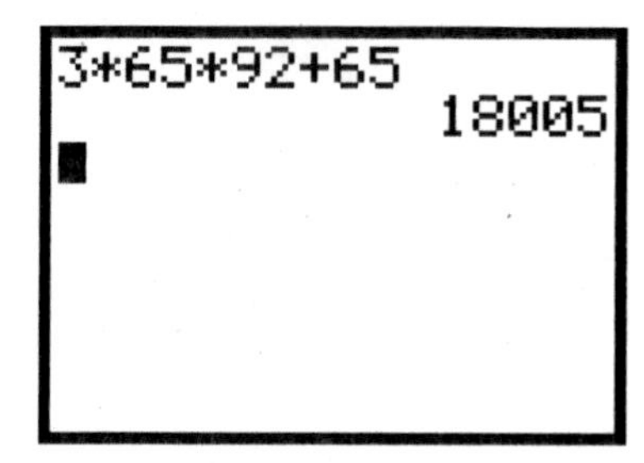

You can recall and edit your entry if necessary. If, for instance, in the expression above you pressed 8 instead of 9, first press [2nd] [ENTRY] to return to the last entry. (ENTRY is the second operation associated with the [ENTER] key.) Then use the [◁] key to move the cursor to 8 and press 9 to overwrite it. If you forgot to type the 2, move the cursor to the plus sign; then press [2nd] [INS] 2 to insert the 2 before the plus sign. (INS is the second operation associated with the [DEL] key.) You can continue to insert symbols immediately after the first insertion without pressing [2nd] [INS] again. If you typed 31 instead of 3, move the cursor to 1 and press [DEL]. This will delete the 1. If you notice that an entry needs to be edited before you press [ENTER] to perform the computation, the editing can be done directly without recalling the entry.

The keystrokes [2nd] [ENTRY] can be used repeatedly to recall entries preceding the last one. Pressing [2nd] [ENTRY] twice, for example, will recall the next to last entry. Using these keystrokes a third time recalls the third to last entry and so on. The number of entries that can be recalled depends on the amount of storage they occupy in the calculator's memory.

USING A MENU

A menu is a list of options that appears when a key is pressed. Thus, multiple options, and sometimes multiple menus, may be accessed by pressing one key. For example, the following screen appears when the [MATH] key is pressed. We see the names of four submenus, MATH, NUM, CPX, and PRB as well as the options in the MATH submenu.

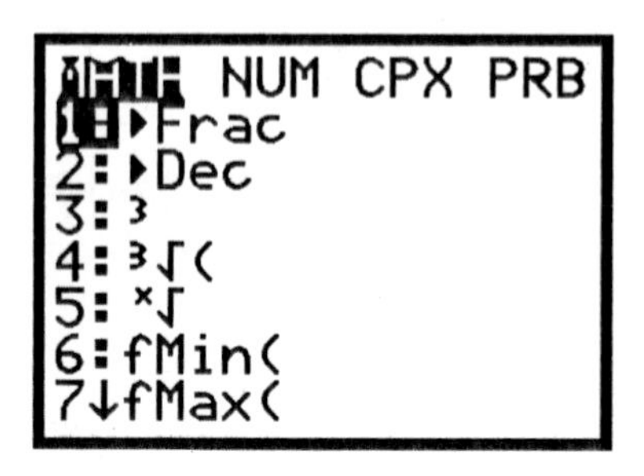

We can copy an item from a menu to the home screen either by using the up or down arrow key to highlight its number and then pressing [ENTER] or by simply pressing the number of the item. The down-arrow beside item 7 in the menu above indicates that there are additional items in the menu. Use the [▽] key to scroll down to them.

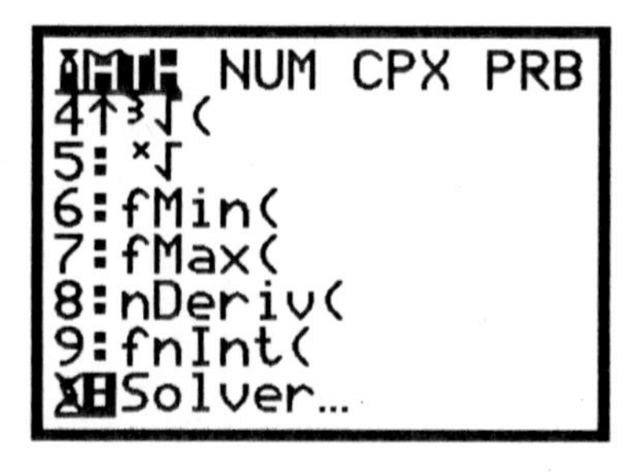

The next example involves choosing an option from a menu.

Section 1.3, Example 12 Use a graphing calculator to find fraction notation for $\frac{2}{15} + \frac{7}{12}$.

Enter $\frac{2}{15} + \frac{7}{12}$ by pressing 2 [÷] 1 5 [+] 7 [÷] 1 2. Now press [ENTER] to find decimal notation for the sum. To convert this to fraction notation we select the ▹Frac feature from the MATH submenu of the MATH menu. Press [MATH] 1 [ENTER] or [MATH] [ENTER] [ENTER]. These keystrokes recall the previous answer and then convert it to fraction notation. Note that this conversion must be done immediately after the calculation is performed in order to have the result of the calculation available for the conversion.

```
2/15+7/12
          .7166666667
Ans▸Frac
                43/60
```

We can also find fraction notation for this sum in one step by selecting the ▹Frac operation before the sum is computed. When this is done, decimal notation does not appear on the screen. Press 2 [÷] 1 5 [+] 7 [÷] 1 2 [MATH] 1 [ENTER] or 2 [÷] 1 5 [+] 7 [÷] 1 2 [MATH] [ENTER] [ENTER].

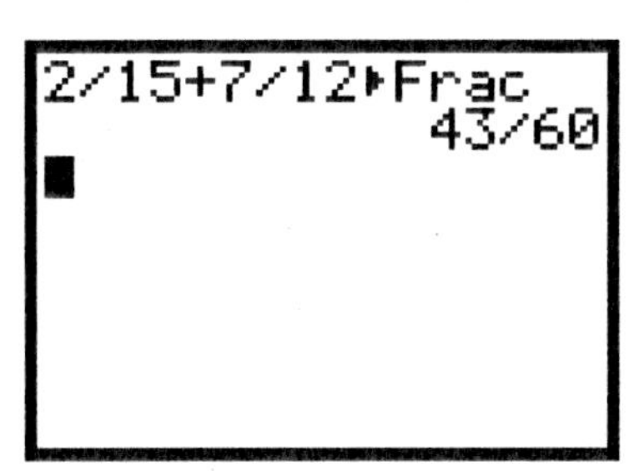

SQUARE ROOTS

Section 1.4, Example 5 Graph the real number $\sqrt{3}$ on a number line.

We can use the calculator to find a decimal approximation for $\sqrt{3}$. Press [2nd] [$\sqrt{\ }$] 3 [)] [ENTER]. Note that the calculator supplies a left parenthesis along with the radical symbol. Although it is not necessary to supply a right parenthesis in this case, we will do so for completeness.

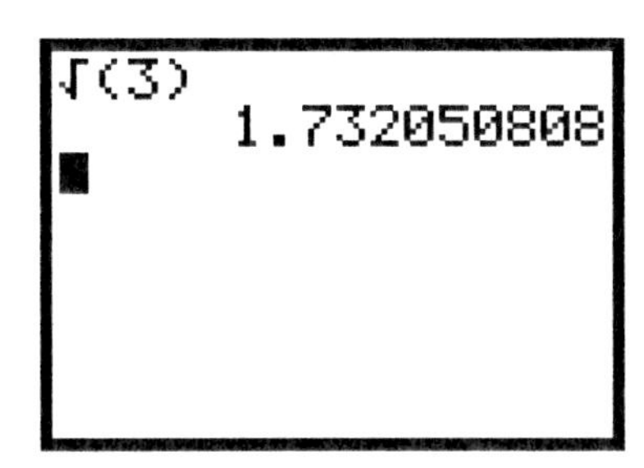

This approximation can now be used to locate $\sqrt{3}$ on the number line. The graph appears on page 33 of the text.

NEGATIVE NUMBERS AND ABSOLUTE VALUE

On the TI-83, TI-83 Plus, and TI-84 Plus, $|x|$ is written abs(X). Absolute value notation is item 1 on the MATH NUM menu.

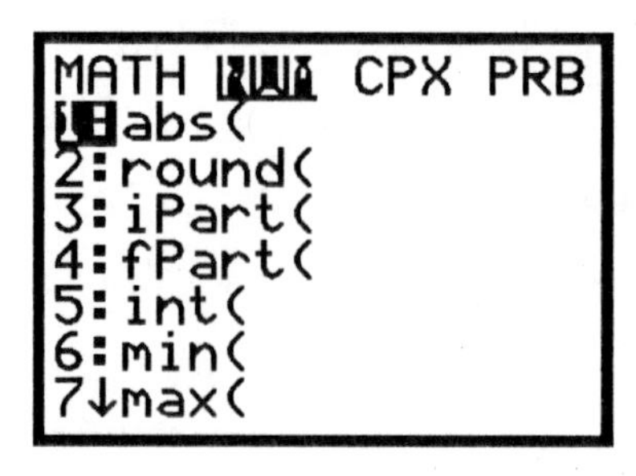

Section 1.4, Example 8 Find each absolute value: **(a)** $|-3|$; **(b)** $|7.2|$; **(c)** $|0|$.

When entering $|-3|$, keep in mind that the [(−)] key in the bottom row of the keypad must be used to enter a negative number on the graphing calculator whereas the [−] key in the right-hand column of the keypad is used to enter the operation of subtraction. Press [MATH] [▷] [ENTER] [(−)] 3 [)] [ENTER]. To find $|7.2|$ use the keystrokes above, replacing [(−)] 3 with 7 [.] 2 and to find $|0|$ replace [(−)] 3 with 0. Note that the calculator supplies a left parenthesis along with the absolute value notation. We supply the right parenthesis to close the absolute-value expression.

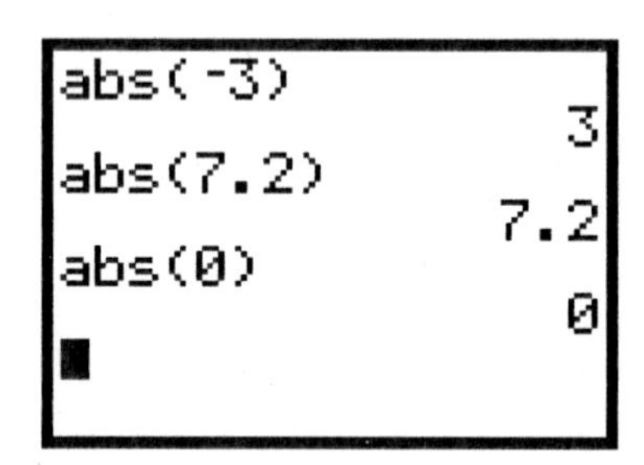

After finding $|-3|$, we could also find $|7.2|$ by recalling the previous entry ($|-3|$) and then editing that entry to replace -3 with 7.2. We could then recall the entry $|7.2|$ and edit it to find $|0|$. (See page 4 of this manual for a discussion of editing entries.)

Instead of pressing [MATH] [▷] [ENTER] to access "abs(" and copy it to the home screen, we could have pressed [MATH] [▷] 1 since "abs(" is item 1 on the MATH NUM menu. Absolute value notation can also be found as the first item in the CATALOG and copied to the home screen. To do this press [2nd] [CATALOG] [ENTER]. (CATALOG is the second operation associated with the 0 numeric key.)

ORDER OF OPERATIONS; EXPONENTS AND GROUPING SYMBOLS

The TI-83, TI-83 Plus, and Ti-84 Plus follow the rules for order of operations.

Section 1.8, Example 8 Calculate: $\dfrac{12(9-7)+4\cdot 5}{2^4+3^2}$.

The fraction bar must be replaced with a set of parentheses around the entire numerator and another set of parentheses around the entire denominator when this expression is entered in the calculator. To enter an exponential expression, first enter the base, then use the [∧] key followed by the exponent. If the exponent is 2, the keystrokes [∧] 2 can be replaced with the single keystroke [x^2].

To enter the expression above and express the result in fraction notation, first press [(] 1 2 [(] 9 [−] 7 [)] [+] 4 [×] 5 [)] [÷] [(] 2 [∧] 4 [+] 3 [x^2] [)]. Remember to use the [−] key for subtraction rather than the [(−)] key, which is used to enter negative numbers. Then press [MATH] 1 or [MATH] [ENTER] to choose the Frac option from the MATH MATH menu. Finally, press [ENTER] to see the result.

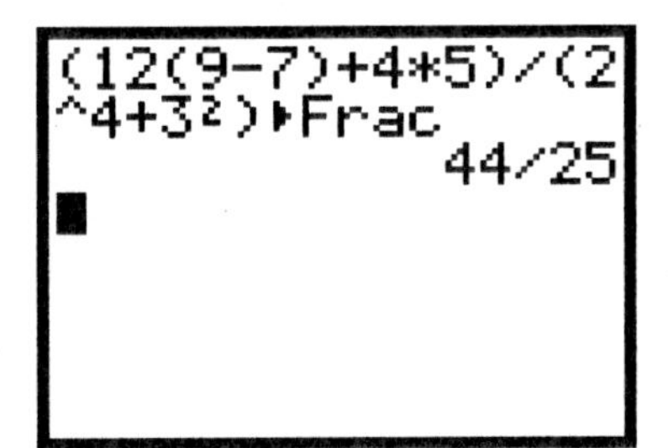

Chapter 2
Equations, Inequalities, and Problem Solving

EVALUATING EXPRESSIONS

Section 2.1, Example 3 Use a calculator to determine whether each of the following is a solution of $2x - 5 = -7$; $3, -1$.

For a given value of x, if the value of the expression on the left side of the equation, $2x - 5$, is the same as the number on the right side of the equation, -7, then the given value of x is a solution of the equation. We can evaluate the expression on the left side for each value of x directly by substituting that value for x as described on page 3 of this manual. We could also evaluate the expression $2x - 5$ by storing a value of x in the calculator first and then entering the expression in terms of the variable.

To store 3 as x, press 3 [STO▷] [X, T, θ, n] [ENTER]. Then enter the expression $2x - 5$ by pressing 2 [X, T, θ, n] [−] 5. Finally, press [ENTER] to evaluate the expression. We see that $2x - 5$ is 1 when $x = 3$, so 3 is not a solution of the equation $2x - 5 = -7$.

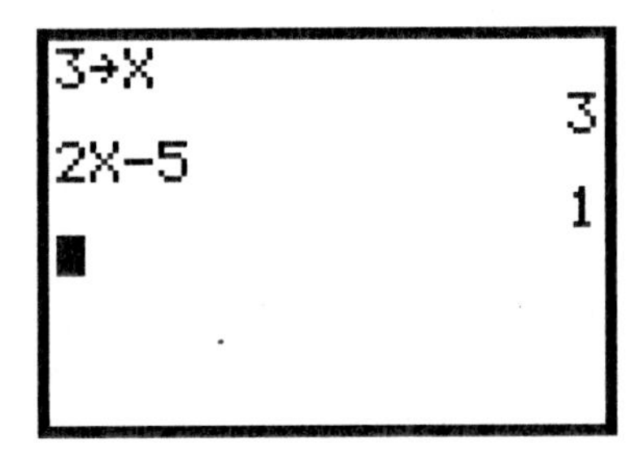

Next store -1 as x by pressing [(−)] 1 [STO▷] [X, T, θ, n] [ENTER]. Then press [2nd] [ENTRY] [2nd] [ENTRY] to recall the expression $2x - 5$ to the screen. Now press [ENTER]. We see that $2x - 5$ is -7 for $x = -1$, so -1 is a solution of the equation $2x - 5 = -7$.

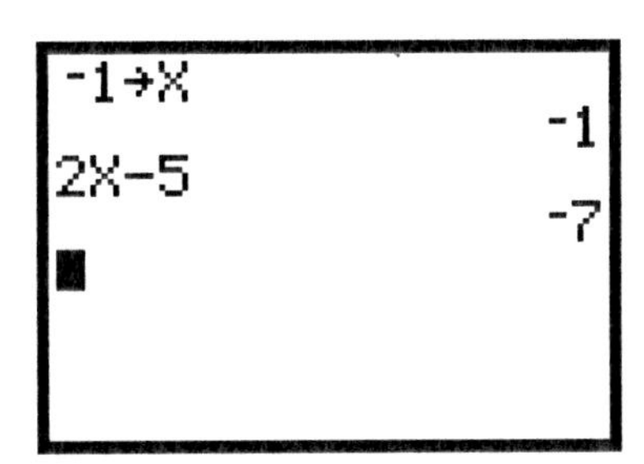

EDITING ENTRIES

In **Section 2.1** the procedure for editing entries is discussed on page 75. This procedure is described on page 4 of this manual.

EVALUATING FORMULAS; THE TABLE FEATURE: ASK MODE

Section 2.3, Example 2 Use the formula $B = 30a$, described in Example 1, to determine the minimum furnace output for well-insulated houses containing 800 ft^2, 1500 ft^2, 2400 ft^2, and 3600 ft^2.

First we replace B with y and a with x and enter the formula $y = 30x$ on the equation-editor screen as equation y_1. Press [Y =] to access this screen. If any of Plot 1, Plot 2, and Plot 3 is turned on (highlighted), turn it off by using the arrow keys to move the blinking cursor over the plot name and pressing ENTER. If there is currently an expression displayed for y_1, clear it by positioning the cursor beside "Y_1 =" and pressing CLEAR. Do the same for expressions that appear on all other lines by using ▽ to move to a line and then pressing CLEAR. Then use △ or ▽ to move the cursor to the top line beside "Y_1 =." Now press 3 0 X, T, Θ, n to enter the right-hand side of the equation on the "Y =" screen.

```
Plot1 Plot2 Plot3
\Y1=30X
\Y2=
\Y3=
\Y4=
\Y5=
\Y6=
\Y7=
```

For an equation entered in the equation-editor screen, a table of x-and y-values can be displayed. We will use a table to evaluate the formula for the given values.

Once the formula is entered, press 2nd TBLSET to display the table set-up screen. (TBLSET is the second function associated with the WINDOW key.) You can choose to supply the x-values yourself or you can set the calculator to supply them. To choose the x-values yourself, set "Indpnt" to "Ask" by using the ▽ and ▷ keys to position the cursor over "Ask" and then pressing ENTER. "Depend" should be set to "Auto." In Ask mode the graphing calculator disregards the settings of TblStart and ΔTbl.

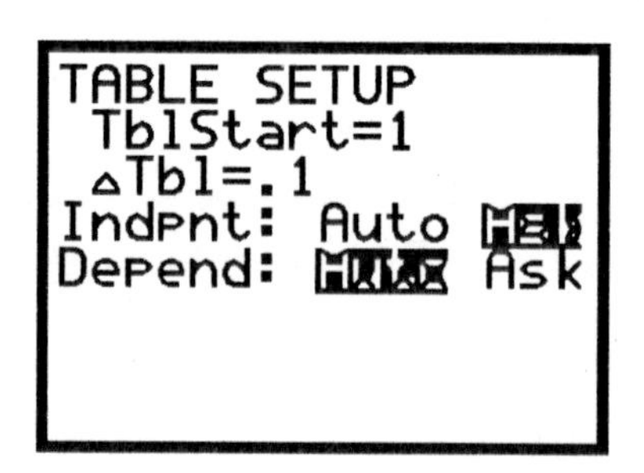

Now press 2nd TABLE to view the table. (TABLE is the second operation associated with the GRAPH key.) Values for x can be entered in the X-column of the table and the corresponding values of y_1 will be displayed in the Y_1-column. To enter 800, 1500, 2400, and 3600, press 8 0 0 ENTER 1 5 0 0 ENTER 2 4 0 0 ENTER 3 6 0 0 ENTER. The down arrow key ▽ can be pressed instead of ENTER if desired.

X	Y1
800	24000
1500	45000
2400	72000
3600	108000

X=

We see that $y_1 = 24,000$ when $x = 800$, $y_1 = 45,000$ when $x = 1500$, $y_1 = 72,000$ when $x = 2400$, and $y_1 = 108,000$ when $x = 3600$, so the furnace outputs for 800 ft^2, 1500 ft^2, 2400 ft^2, and 3600 ft^2 are 24,000 Btu's, 45,000 Btu's, 72,000 Btu's, and 108,000 Btu's, respectively.

THE TABLE FEATURE: AUTO MODE

Section 2.4, Example 8 Village Stationers wants the customer service team to be able to look up the cost c of merchandise being returned when only the total amount paid T (including tax) is shown on the receipts. Use the formula $c = T/1.05$, developed in Example 7, to create a table of values showing cost given a total amount paid. Assume that the least possible sale is \$0.21.

We will use a graphing calculator to create the table of values. To enter the formula, we first replace c with y and T with x. Then we enter the formula on the equation-editor screen as $y = x/1.05$. Be sure that the Plots are turned off and that any previous entries are cleared. (See pages 9 and 10 of this manual for the procedure for turning off the Plots and clearing equations.)

```
Plot1 Plot2 Plot3
\Y1=X/1.05
\Y2=
\Y3=
\Y4=
\Y5=
\Y6=
\Y7=
```

We will create a table in which the calculator supplies the x-values beginning with a value we specify and continuing by adding a value we specify to the preceding value for x. We will begin with an x-value of 0.21 (corresponding to \$0.21) and choose successive increases of 0.01 (corresponding to \$0.01). To do this, first press [2nd] [TBLSET] to access the TABLE SETUP window. If "Indpnt" is set to "Auto," the calculator will supply values for x, beginning with the value specified as TblStart and continuing by adding the value of ΔTbl to the preceding value for x. Press [.] 2 1 [▽] [.] 0 1 to select a beginning x-value of 0.21 and an increment of 0.01. The "Indpnt" and "Depend" settings should both be "Auto." If either is not, use the [▽] key to position the blinking cursor over "Auto" on that line and then press [ENTER]. To display the table press [2nd] [TABLE].

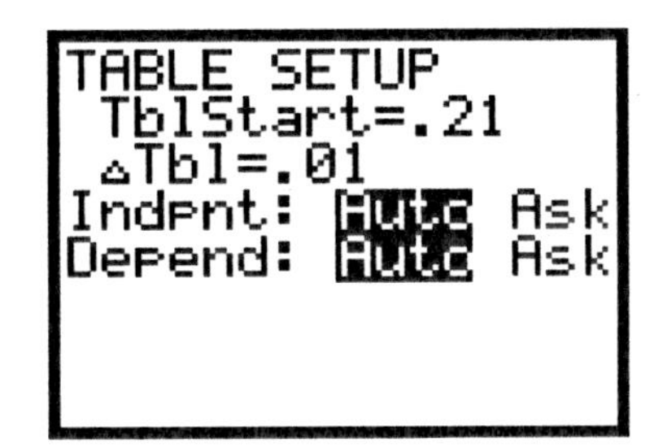

X	Y1
.21	.2
.22	.20952
.23	.21905
.24	.22857
.25	.2381
.26	.24762
.27	.25714

X=.21

We can change the number of decimal places that the graphing calculator will display to 2 so that the costs are rounded to the nearest cent. To do this, press [MODE] to access the MODE screen. Then press [▽] to move the cursor to the second line and press the [▷] key three times to highlight 2. (This assumes that "Float" was previously selected.) Finally, press [ENTER] to select two decimal places. To return to the table, press [2nd] [TABLE].

```
Normal Sci Eng
Float 0123456789
Radian Degree
Func Par Pol Seq
Connected Dot
Sequential Simul
Real a+bi re^θi
Full Horiz G-T
```

X	Y1
.21	.20
.22	.21
.23	.22
.24	.23
.25	.24
.26	.25
.27	.26

X=.21

Use the ▽ and △ keys to scroll through the table.

Before proceeding, return to the MODE screen and reselect "Float." This allows the number of decimal places to vary, or float, according to the computation being performed.

In **Section 2.5, Example 2**, two equations are entered on the equation-editor screen. The first equation, $y_1 = x + 1$, can be entered as described on pages 9 and 10 of this manual. Then, to enter $y_2 = x + (x + 1)$, first press either ▽ or ENTER to position the cursor beside "Y_2 =." Now enter the right-hand side of the equation.

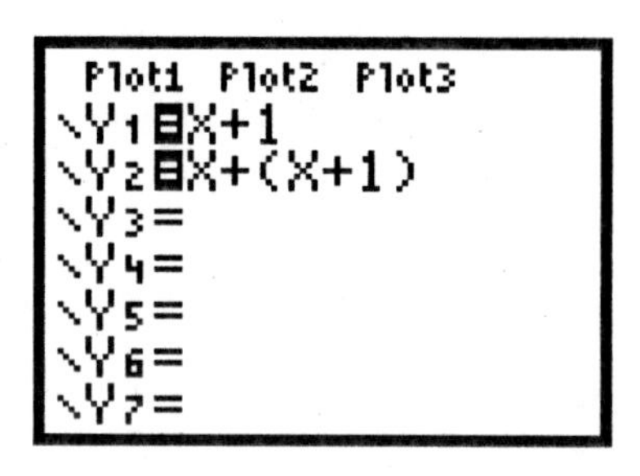

Chapter 3
Graphs and Linear Equations

If you selected 2 for the number of decimal places in Section 2.4, Example 8, and have not yet returned to the MODE screen to reselect "Float," do so now. In Float mode the number of decimal places varies, or floats, according to the computation being performed.

SETTING THE VIEWING WINDOW

The viewing window is the portion of the coordinate plane that appears on the graphing calculator's screen. It is defined by the minimum and maximum values of x and y: Xmin, Xmax, Ymin, and Ymax. The notation [Xmin, Xmax, Ymin, Ymax] is used in the text to represent these window settings or dimensions. For example, $[-12, 12, -8, 8]$ denotes a window that displays the portion of the x-axis from -12 to 12 and the portion of the y-axis from -8 to 8. In addition, the distance between tick marks on the axes is defined by the settings Xscl and Yscl. In this manual Xscl and Yscl will be assumed to be 1 unless noted otherwise. The setting Xres sets the pixel resolution. We usually select Xres = 1. The window corresponding to the settings $[-20, 30, -12, 20]$, Xscl = 5, Yscl = 2, Xres = 1, is shown below. Note that all of the entries on the equation-editor screen have been cleared and the plots have been turned off in order to show this window. (See pages 9 and 10 of this manual for the procedure for clearing entries and turning off the plots.)

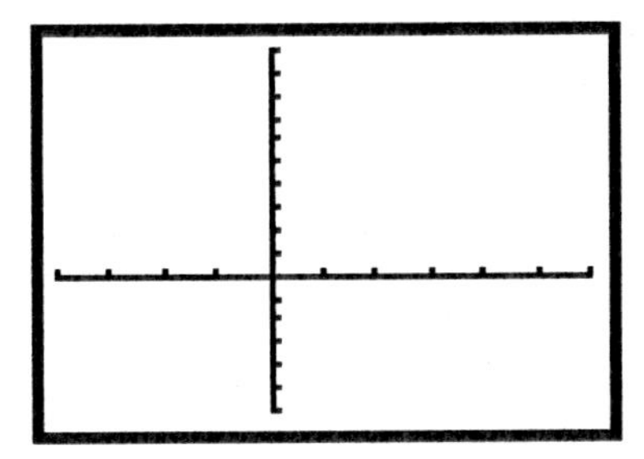

Press the WINDOW key on the top row of the keypad to display the current window settings on your calculator. The standard settings are shown below.

```
WINDOW
 Xmin=-10
 Xmax=10
 Xscl=1
 Ymin=-10
 Ymax=10
 Yscl=1
 Xres=1
```

Section 3.1, Example 5 Set up a $[-100, 100, -5, 5]$ viewing window on a graphing calculator, choosing appropriate scales for the axes.

To change a setting, position the cursor beside the setting you wish to change and enter the new value. We will enter the given settings and let Xscl = 10 and Yscl =1. The choice of scales for the axes may vary. We select scaling that will allow some space between tick marks. If a scale that is too small is chosen, the tick marks will blend and blur. To change the settings to $[-100, 100, -5, 5]$, Xscl = 10, Yscl = 1, press WINDOW (−) 1 0 0 ENTER 1 0 0 ENTER 1 0 ENTER (−) 5 ENTER 5 ENTER 1 ENTER. The ▽ key may be used instead of ENTER after typing each window setting. To see the window with these settings, press the GRAPH key on the top row of the keypad.

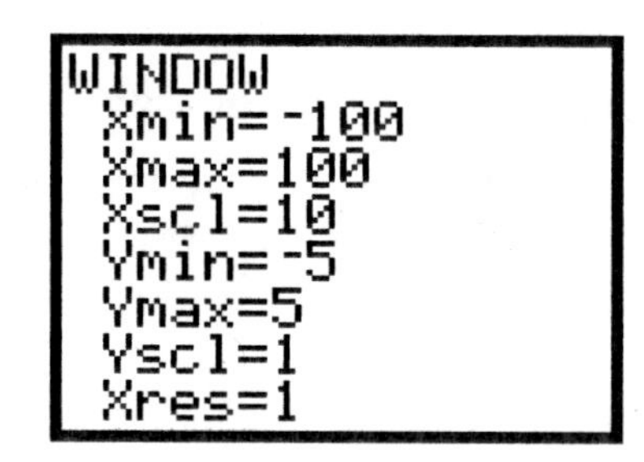

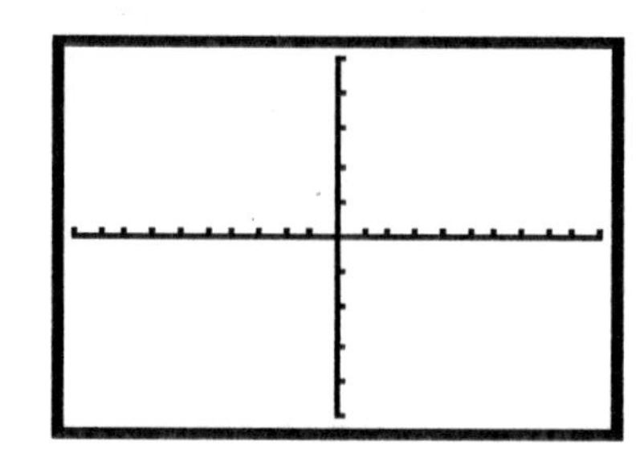

QUICK TIP: To return quickly to the standard window setting $[-10, 10, -10, 10]$, Xscl = 1, Yscl = 1, press ZOOM 6.

GRAPHING EQUATIONS

After entering an equation and setting a viewing window, you can view the graph of an equation.

Section 3.2, Example 4 Graph $y = 2x$ using a graphing calculator.

Press Y = to access the equation-editor screen. Turn off the plots and clear any previous entries. (See pages 9 and 10 of this manual.) Then use △ or ▽ to move the cursor to the top line beside "Y_1 =." Now press 2 X, T, Θ, n to enter the right-hand side of the equation on the "Y =" screen.

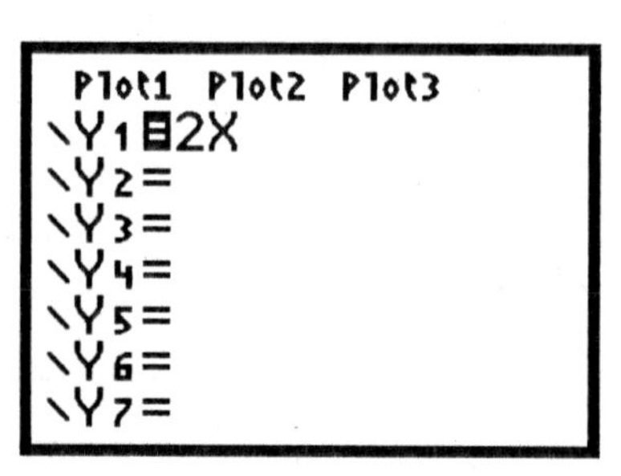

The standard $[-10, 10, -10, 10]$ window is a good choice for this graph. Either enter these dimensions in the WINDOW screen and then press GRAPH to see the graph or simply press ZOOM 6 to select the standard window and see the graph.

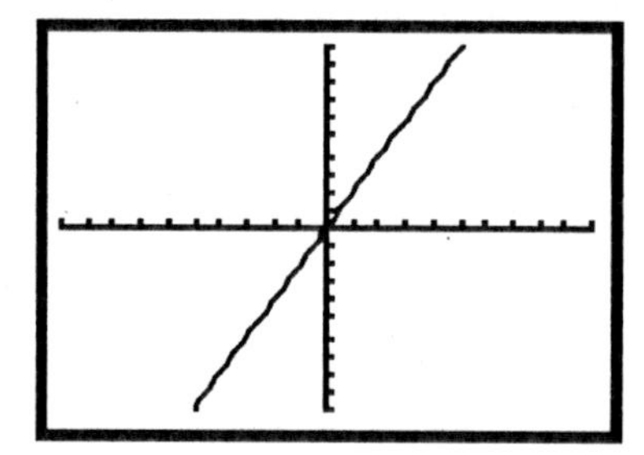

SOLVING EQUATIONS GRAPHICALLY; THE INTERSECTION METHOD

We can use the Intersect feature from the CALC menu to solve equations.

Section 3.3, Example 2 Solve using a graphing calculator: $-\frac{3}{4}x + 6 = 2x - 1$.

On the equation-editor screen, clear any existing entries and then enter $y_1 = -(3/4)x + 6$ and $y_2 = 2x - 1$. Although the parentheses in y_1 are not necessary, they make the equation easier to read on the equation-editor screen. Press [ZOOM] 6 to graph these equations in the standard viewing window. The solution of the equation $-\frac{3}{4}x + 6 = 2x - 1$ is the first coordinate of the point of intersection of the graphs. To use the Intersect feature to find this point, first press [2nd] [CALC] 5 to select Intersect from the CALC menu. (CALC is the second operation associated with the [TRACE] key on the top row of the keypad.) The query "First curve?" appears at the bottom of the screen. The blinking cursor is positioned on the graph of y_1. This is indicated by the notation "$Y_1 = -\frac{3}{4}x + 6$" in the upper left-hand corner of the screen. Press [ENTER] to indicate that this is the first curve involved in the intersection. Next the query "Second curve?" appears at the bottom of the screen. The blinking cursor is now positioned on the graph of y_2 and the notation "$Y_2 = 2x - 1$" should appear in the top left-hand corner of the screen. Press [ENTER] to indicate that this is the second curve. We identify the curves for the calculator since we could have as many as ten graphs on the screen at once. After we identify the second curve, the query "Guess?" appears at the bottom of the screen. Use the right and left arrow keys to move the blinking cursor close to the point of intersection of the graphs. This provides the calculator with a guess as to the coordinates of this point. We do this since some pairs of curves can have more than one point of intersection. When the cursor is positioned, press [ENTER] a third time. Now the coordinates of the point of intersection appear at the bottom of the screen. We see that $x = 2.5454545$, so the solution of the equation is 2.5454545.

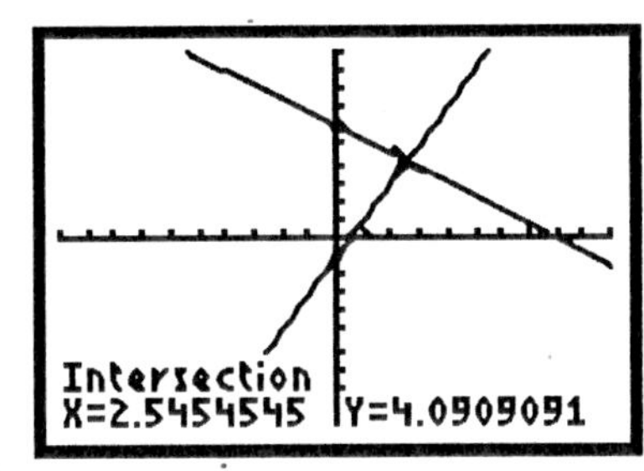

We can check the solution by evaluating both sides of the equation $-\frac{3}{4}x + 6 = 2x - 1$ for this value of x. The first coordinate of the point of intersection has automatically been stored as x in the calculator, so we evaluate y_1 and y_2 for this value of x. First press [2nd] [QUIT] to go to the home screen. Then to evaluate Y_1 press [VARS] [▷] 1 1 [ENTER]. To evaluate Y_2 press [VARS] [▷] 1 2 [ENTER]. We see that Y_1 and Y_2 have the same value when $X = 2.5454545$, so the solution checks.

```
Y1
        4.090909091
Y2
        4.090909091
```

Note that although the procedure above verifies that 2.5454545 is the solution, this number is actually an approximation of the solution. In some cases the calculator will give an exact solution. Since the x-coordinate of the point of intersection is stored in the calculator as X, we can find an exact solution by converting X to fraction notation. This can be done from the home screen by pressing [X ,T, Θ, n] [MATH] 1 [ENTER] or [X ,T, Θ, n] [MATH] [ENTER] [ENTER]. We see that the exact solution is $\frac{28}{11}$.

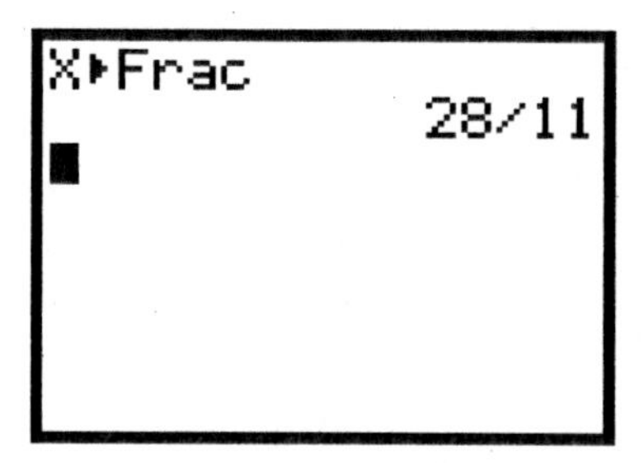

SOLVING EQUATIONS GRAPHICALLY; THE ZERO METHOD

As an alternative to the Intersection method, we can use the Zero feature from the CALC menu to solve equations.

Section 3.3, Example 7 Solve $3 - 8x = 5 - 7x$ using the zero method.

First we get zero on one side of the equation:

$$3 - 8x = 5 - 7x$$
$$-2 - 8x = -7x$$
$$-2 - x = 0.$$

Then we graph $y = -2 - x$. We will use the standard window.

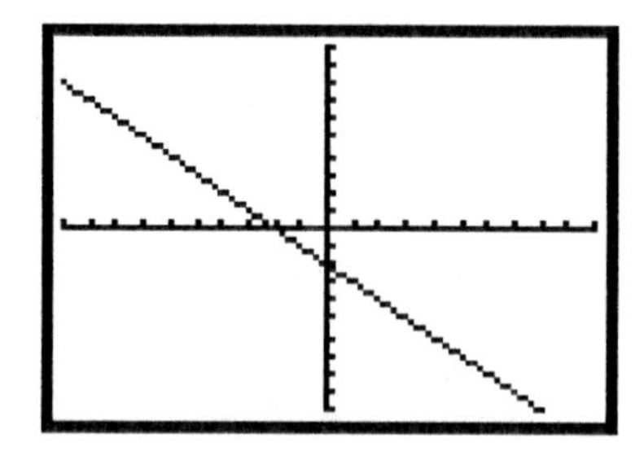

The first coordinate of the x-intercept of the graph is the zero of $y = -2 - x$ and thus is the solution of the original equation. Press [2nd] [CALC] 2 to select Zero from the CALC menu. We are prompted to select a left bound. This means that we must choose an x-value that is to the left of the first coordinate of the x-intercept. This can be done by using the left and right arrow keys to move the cursor to a point on the graph that is to the left of the intercept or by keying in an x-value that is less than the first coordinate of the intercept.

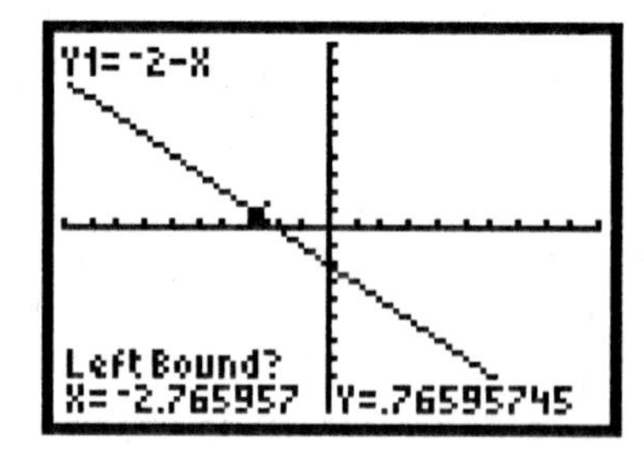

Once this is done, press [ENTER]. Now we are prompted to select a right bound that is to the right of the x-intercept. Again we can use the arrow keys or key in a value.

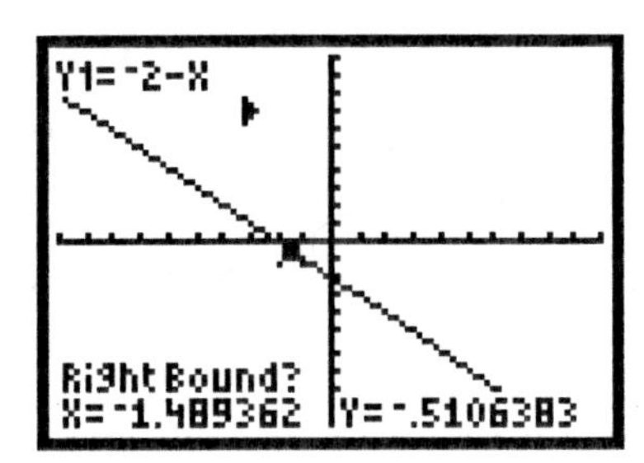

Press ENTER again. Finally, we are prompted to make a guess as to the value of the zero. Move the cursor to a point close to the zero or key in a value.

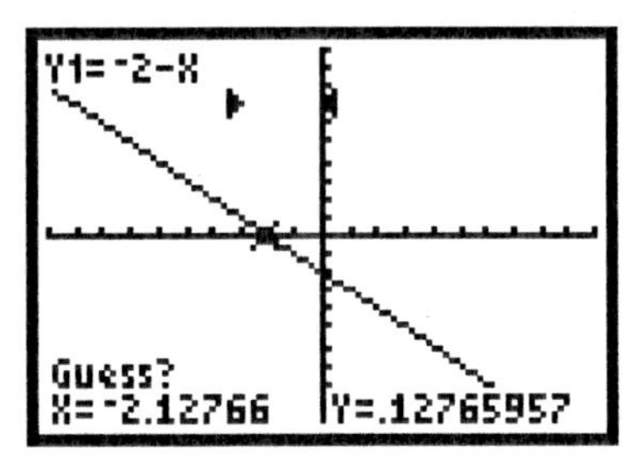

Press ENTER a third time. We see that $y = 0$ when $x = -2$, so -2 is the zero of $y = -2 - x$ and is thus the solution of the original equation, $3 - 8x = 5 - 7x$.

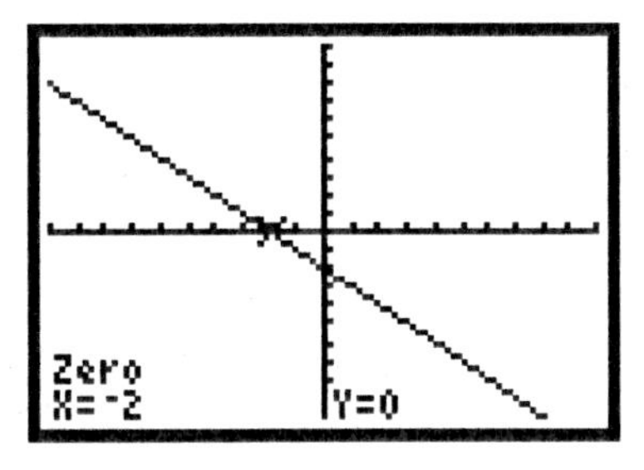

SQUARING THE VIEWING WINDOW

Section 3.7, Example 9 Determine whether the lines given by the equations $3x - y = 7$ and $x + 3y = 1$ are perpendicular, and check by graphing.

In the text each equation is solved for y in order to determine the slopes of the lines. We have $y = 3x - 7$ and $y = -\frac{1}{3}x + \frac{1}{3}$. Since $3\left(-\frac{1}{3}\right) = -1$, we know that the lines are perpendicular. To check this, we graph $Y_1 = 3x - 7$ and $Y_2 = -\frac{1}{3}x + \frac{1}{3}$. The graphs are shown on the right below in the standard viewing window.

```
WINDOW
 Xmin=-10
 Xmax=10
 Xscl=1
 Ymin=-10
 Ymax=10
 Yscl=1
 Xres=1
```

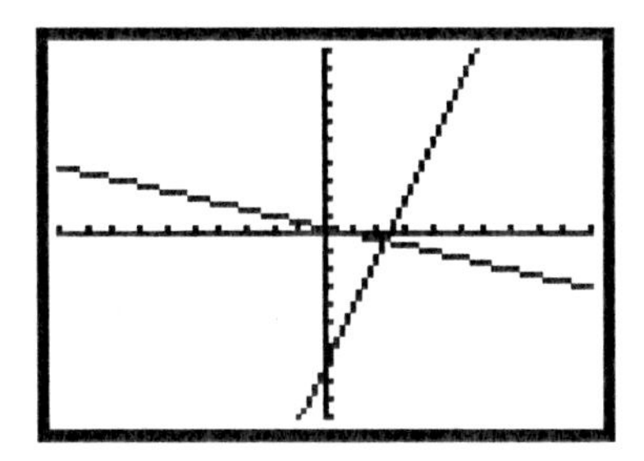

Note that the graphs do not appear to be perpendicular. This is due to the fact that, in the standard window, the distance between tick marks on the y-axis is about 2/3 the distance between tick marks on the x-axis. It is often desirable to choose window dimensions for which these distances are the same, creating a "square" window. On the TI-83, TI-83 Plus, and TI-84 Plus, any window in which the ratio of the length of the y-axis to the length of the x-axis is $2/3$ will produce this effect.

This can be accomplished by selecting dimensions for which Ymax − Ymin $= \frac{2}{3}$(Xmax − Xmin). For example, the windows $[-12, 12, -8, 8]$ and $[-6, 6, -4, 4]$ are square. When we change the window dimensions to $[-12, 12, -8, 8]$ and press GRAPH, the graphs now appear to be perpendicular as shown on the right below. Instead of entering window dimensions, we could also press ZOOM 5 and the calculator will select a square window.

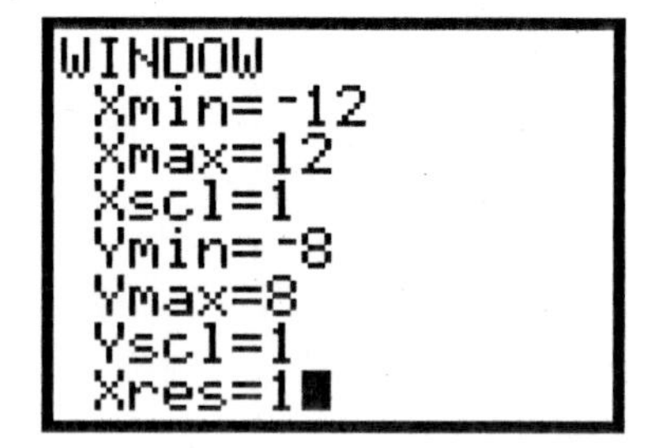

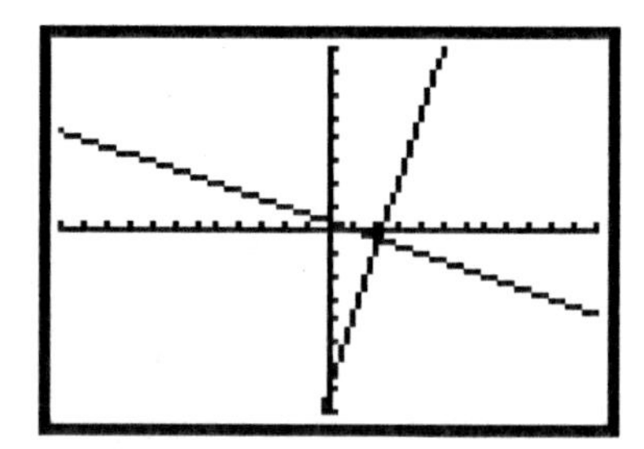

Chapter 4
Systems of Equations and Problem Solving

SOLVING SYSTEMS OF EQUATIONS BY GRAPHING

The INTERSECT option from the CALC menu can be used to solve a system of two equations in two variables. The procedure for finding the point of intersection of the graphs of two equations is described on page 15 of this manual.

GRAPHING INEQUALITIES IN TWO VARIABLES

The solution set of an inequality in two variables can be graphed on the TI-83, the TI-83 Plus, and the TI-84 Plus.

Section 4.5, Example 5 Graph $2x + 3y \leq 6$ using a graphing calculator.

Solving for y, we have $y \leq -\frac{2}{3}x + 2$. Then the boundary line is $y = -\frac{2}{3}x + 2$. We will enter this as Y_1. Press [Y =] to go to the equation-editor screen. If there is currently an entry for Y_1, clear it. Also clear any other equations that are entered. Now enter $y_1 = (-2/3)x + 2$. Since the inequality is in the form $y \leq mx + b$, we shade the half-plane below the graph of y_1.

Seven graph styles can be selected on the equation-editor screen. The "shade below" GraphStyle can be used to shade the region below the graph of the boundary line. To select this option, move the cursor to the GraphStyle icon to the left of Y_1 and press [ENTER] repeatedly until the symbol indicating the "shade below" GraphStyle appears as shown below. If the "line" GraphStyle was previously selected, the "Shade below" icon will appear after [ENTER] is pressed three times. (To shade above a line we would press [ENTER] until the "shade above" GraphStyle symbol appears.) Then press [ZOOM] 6 to see the graph of the inequality in the standard viewing window.

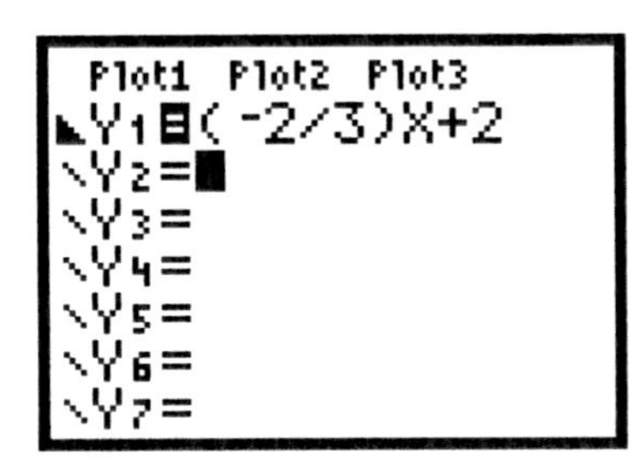

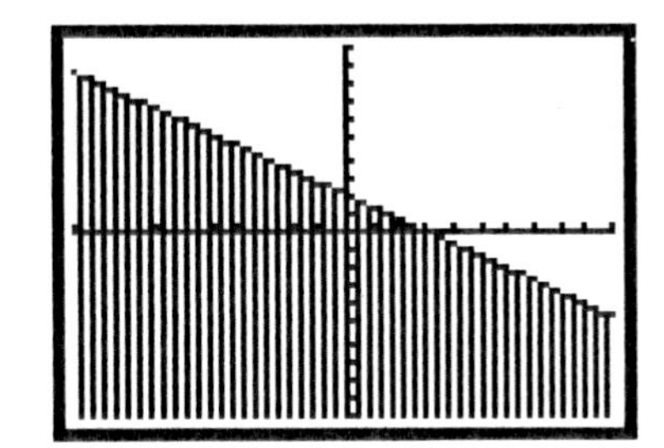

SYSTEMS OF LINEAR INEQUALITIES

We can graph systems of inequalities by shading the solution set of each inequality in the system with a different pattern. When the "shade above" or "shade below" GraphStyle options are selected the calculator rotates through four shading patterns. Vertical lines shade the first equation, horizontal lines the second, negatively sloping diagonal lines the third, and positively sloping diagonal lines the fourth. These patterns repeat if more than four equations are graphed.

Section 4.6, Example 3 Graph the solutions of the system

$$x - 2y < 0,$$
$$-2x + y > 2.$$

First we solve $x - 2y < 0$ for y, obtaining $y > x/2$. We graph the boundary line $y = x/2$ and, since the inequality can be written in the form $y > mx + b$, we select the "shade above" GraphStyle for this entry. (See Example 5 above for instructions on selecting

GraphStyle icons.) Next we solve $-2x + y > 2$ for y, obtaining $y > 2x + 2$. We see that this inequality can be written in the form $y > mx + b$, so we graph the boundary line $y = 2x + 2$ and shade above it. Trial and error shows that $[-7, 5, -6, 6]$ is a good viewing window for this system of inequalities. The graph shows the solution set of each inequality in the system and the region where they overlap. The region of overlap is the solution set of the system of inequalities. Keep in mind that neither boundary line is part of the solution set.

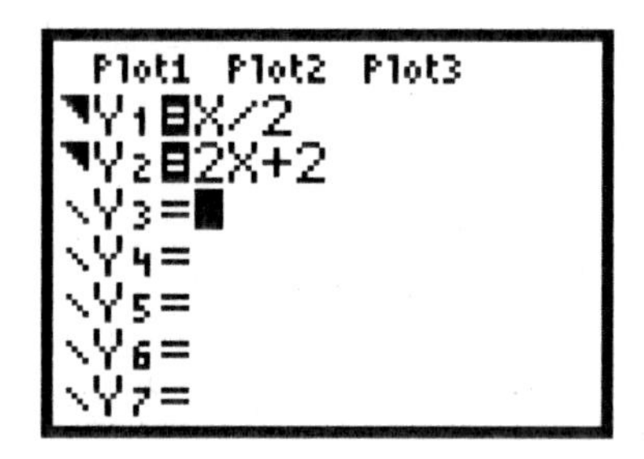

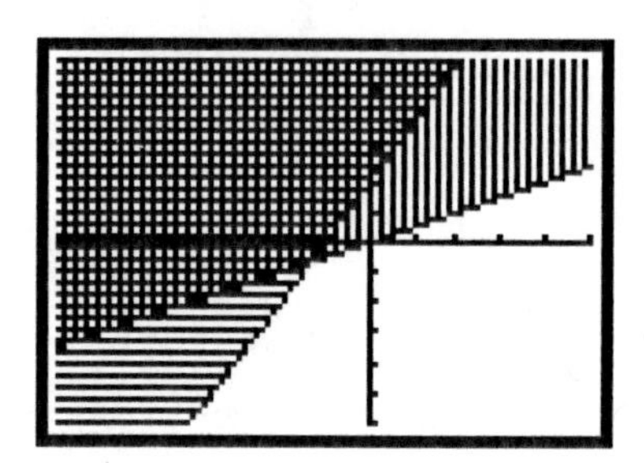

Chapter 5
Polynomials

EVALUATING POLYNOMIALS

Section 5.2, Example 6 Evaluate $-x^2 + 3x + 9$ for $x = -2$.

We can use a graphing calculator to evaluate polynomials in several ways. We can enter the polynomial on the home screen, substituting a value for the variable, as described on page 3 of this manual. We can also use a table set in Ask mode as described on page 10 of this manual.

In addition to the methods mentioned above, we can use a graph to evaluate a polynomial. First we graph $y_1 = -x^2 + 3x + 9$. The window $[-10, 10, -10, 15]$ is a good choice for this graph. Now we can use either the VALUE option from the CALC menu or the TRACE feature to evaluate the polynomial. To select VALUE, press [2nd] [CALC] [ENTER] or [2nd] [CALC] 1. Then supply the desired value of x as indicated by the blinking cursor beside X =. To do this press [(−)] 2 [ENTER]. To use TRACE to evaluate the polynomial after it is graphed, press [TRACE] [(−)] 2 [ENTER]. These keystrokes select the TRACE feature and supply the value -2 for x. In each case we see X = −2, Y = −1 at the bottom of the screen, so the value of $-x^2 + 3x + 9$ is -1 when $x = -2$.

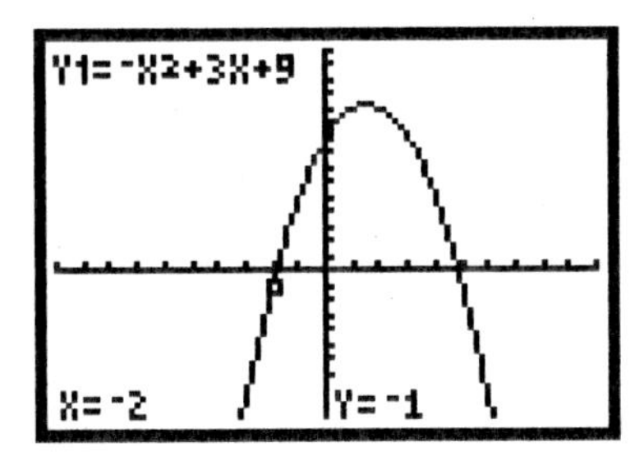

CHECKING OPERATIONS ON POLYNOMIALS

A graphing calculator can be used to check operations on polynomials.

Section 5.3, pages 320 and 321 Several methods for checking whether two algebraic expressions are equivalent are discussed on pages 320 and 321 of the text. We will illustrate these methods by checking the addition $(-3x^3 + 2x - 4) + (4x^3 + 3x^2 + 2) = x^3 + 3x^2 + 2x - 2$.

First we will check this result by comparing the graphs of $Y_1 = (-3x^3 + 2x - 4) + (4x^3 + 3x^2 + 2)$ and $Y_2 = x^3 + 3x^2 + 2x - 2$. This is most easily done when different graph styles are used for the graphs. The **path graph style** can be used, along with the line style, to determine whether graphs coincide. To use graphs to check the given addition, first press [MODE] to determine whether Sequential mode is selected. If it is not, position the blinking cursor over Sequential and then press [ENTER]. Next, on the Y = screen, enter $Y_1 = (-3x^3 + 2x - 4) + (4x^3 + 3x^2 + 2)$ and $Y_2 = x^3 + 3x^2 + 2x - 2$. We will select the line graph style for Y_1 and the path style for Y_2. To select these graph styles use [◁] to position the cursor over the icon to the left of the equation and press [ENTER] repeatedly until the desired style icon appears as shown on the right below.

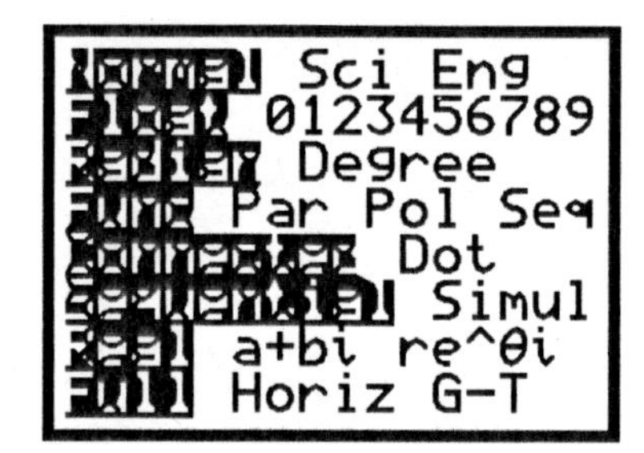

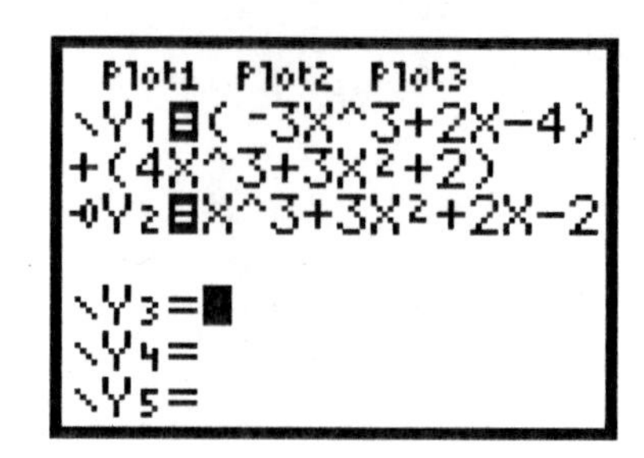

The calculator will graph Y_1 first as a solid line. Then Y_2 will be graphed as the circular cursor traces the leading edge of the graph, allowing us to determine visually whether the graphs coincide. In this case, the graphs appear to coincide, so the factorization is probably correct.

We can also check the addition by **subtracting** the result from the original sum. With Y_1 and Y_2 entered as described above, position the cursor beside "$Y_3 =$" and use the Y-VARS menu to enter $Y_3 = Y_1 - Y_2$ by pressing VARS ▷ 1 1 (−) VARS ▷ 1 2. If the expressions for Y_1 and Y_2 are equivalent, the graph of Y_3 will be $y = 0$, or the x-axis.

Since we are interested only in the values of Y_3, we will deselect Y_1 and Y_2. To do this first position the cursor over the equals sign beside Y_1 and press ENTER. Repeat this for Y_2. Note that the equals signs beside Y_1 and Y_2 are no longer highlighted, indicating that these two equations have been deselected. The graph of an equation that has been deselected will not appear when GRAPH is pressed. A deselected equation can be selected again by positioning the cursor over the equals sign and pressing ENTER. Note that the equals sign is once again highlighted when this is done. Now select the path graph style for Y_3 as described above.

```
Plot1 Plot2 Plot3
\Y1=(-3X^3+2X-4)
+(4X^3+3X²+2)
\Y2=X^3+3X²+2X-2
-0Y3=Y1-Y2
\Y4=
\Y5=
```

Finally, press GRAPH and determine if the graph of Y_3 is traced over the x-axis. Since it is, the sum is correct.

We can use a table of values to **compare values** of Y_1 and Y_2. If the expressions for Y_1 and Y_2 are the same for each given x-value, the result checks. If you deselected Y_1 and Y_2 to check the sum using subtraction above, select them again now as described above. Then look at a table set in Auto mode. (See page 11 of this manual for the procedure for setting up a table in Auto mode.) Since the values of Y_1 and Y_2 are the same for each given x-value, the result checks. Scrolling through the table to look at additional values makes this conclusion more certain.

X	Y1	Y2
-3	-8	-8
-2	-2	-2
-1	-2	-2
0	-2	-2
1	4	4
2	22	22
3	58	58

X= -3

We can also check the sum using a **horizontal split-screen**. The top half of the screen displays the graph and, in this case, we will use the bottom half to display a table of values.

First enter Y_1 and Y_2 as described above. Then, to select the horizontal split-screen option, press MODE to access the Mode screen. Position the cursor over Horiz and press ENTER. Press GRAPH to see the graph in the top half of the split screen, and press 2nd TABLE to see two lines of the table of values for y_1 and y_2 below the graph. We show these equations graphed in the standard window. We show a table with TblStart $= -3$, ΔTbl $= 1$, and Indpnt and Depend both set on Auto. The graphs appear to coincide. In addition, as we scroll through the table, we see that the values of Y_1 and Y_2 are the same for any given x-value, so the sum is probably correct.

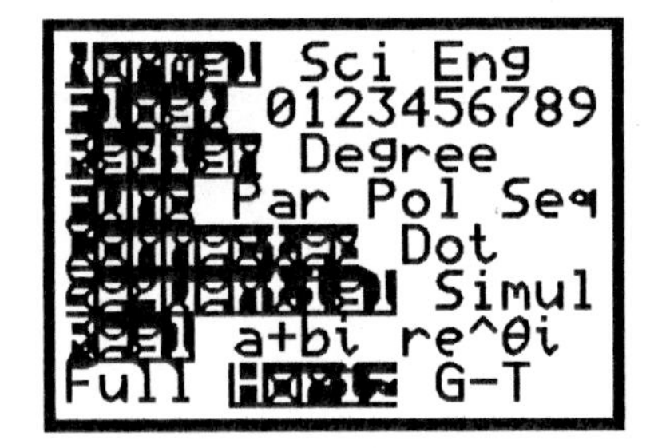

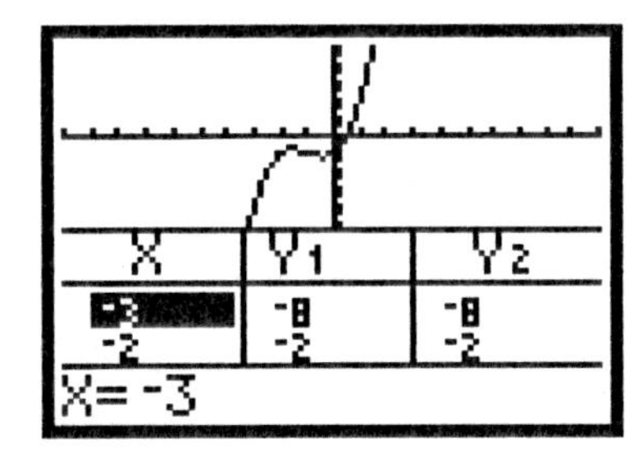

Instead of two lines of the table, the lower half of the split screen can display other information including four lines of the home screen, four lines of the equation-editor screen, or three settings of the Window screen. To change from the table to the home screen, press 2nd QUIT. Display the equation-editor screen by pressing Y =, or the Window screen by pressing WINDOW.

In order to return to a full-screen graph, table, equation-editor, or window screen, return to the Mode screen and select Full.

We can also use a **vertical split screen** to check the addition. First enter Y_1, Y_2, and Y_3 as described above and deselect Y_1 and Y_2. Then press MODE to display the Mode screen, position the cursor over G-T, and press ENTER. Now press GRAPH to see the graph of Y_3 on the left side of the screen and a table of values for Y_3 on the right side. As we did above, we use a table set in Auto mode with TblStart $= -3$ and ΔTbl $= 1$. Since the graph appears to be $y = 0$, or the x-axis, and all of the Y_3-values in the table are 0, we confirm that the result is correct.

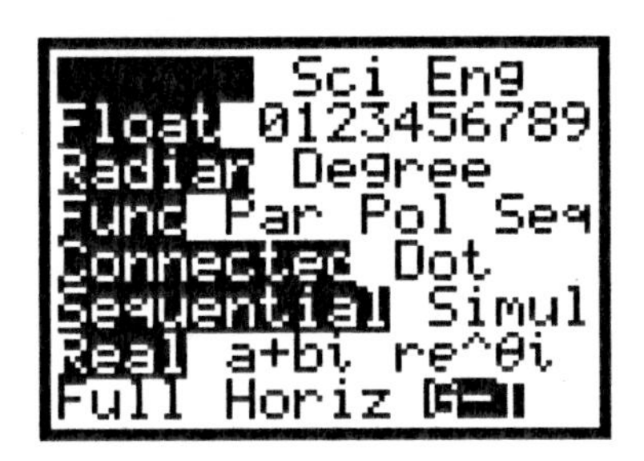

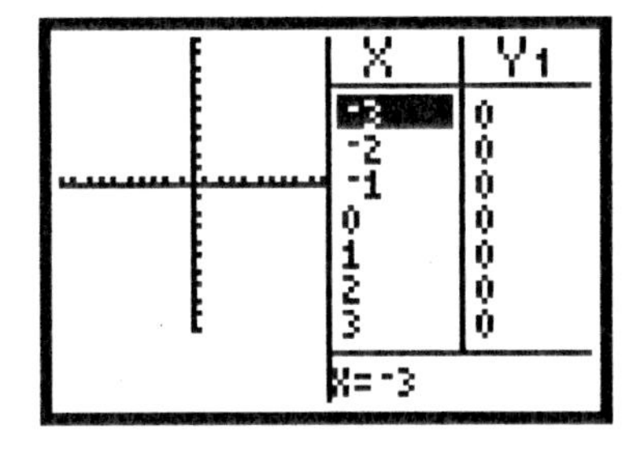

In order to return to full-screen mode, return to the Mode screen and select Full.

EVALUATING POLYNOMIALS IN SEVERAL VARIABLES

Section 5.6, Example 1 Evaluate the polynomial $4 + 3x + xy^2 + 8x^3y^3$ for $x = -2$ and $y = 5$.

To evaluate a polynomial in two or more variables, we substitute numbers for the variables. This can be done by substituting values for x and y directly or by storing the values of the variables in the calculator. The procedure for direct substitution is described on page 3 of this manual.

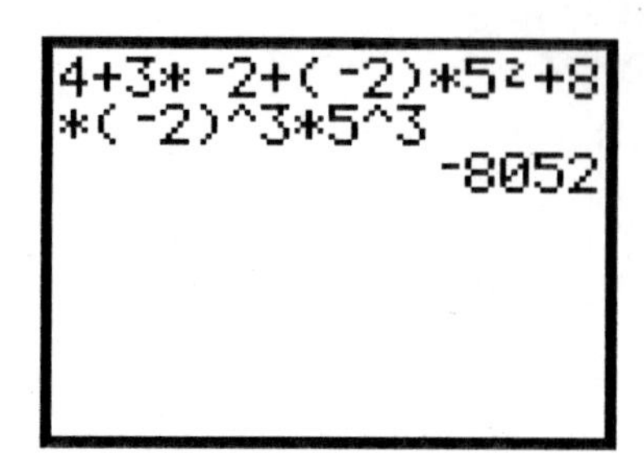

To evaluate the polynomial function by first storing -2 as x and 5 as y, we proceed as follows. To store -2 as x, press (−) 2 STO▷ X, T, Θ, n ENTER. Now store 5 as y. Press 5 STO▷ ALPHA Y ENTER. ALPHA is the green key in the left-hand column of the keypad, and Y is the alpha, or letter, operation associated with the 1 numeric key. Next enter the algebraic expression. Press 4 + 3 X, T, Θ, n + X, T, Θ, n ALPHA Y ∧ 2 + 8 X, T, Θ, n ∧ 3 ALPHA Y ∧ 3. Finally, press ENTER to find the value of the expression.

```
-2→X
                 -2
5→Y
                  5
4+3X+XY²+8X^3Y^3
              -8052
```

SCIENTIFIC NOTATION

To enter a number in scientific notation, first type the decimal portion of the number; then press 2nd EE (EE is the second operation associated with the , key.); finally type the exponent, which can be at most two digits. For example, to enter 1.789×10^{-11} in scientific notation, press 1 . 7 8 9 2nd EE (−) 1 1 ENTER. To enter 6.084×10^{23} in scientific notation, press 6 . 0 8 4 2nd EE 2 3 ENTER. The decimal portion of each number appears before a small E while the exponent follows the E.

```
1.789E-11
          1.789E-11
6.084E23
           6.084E23
```

The calculator can be used to perform computations in scientific notation.

Section 5.8, Example 9(b) Use a graphing calculator to check the computation $(7.2 \times 10^{-7}) \div (8.0 \times 10^{6}) = 9.0 \times 10^{-14}$.

We enter the computation in scientific notation. Press 7 . 2 2nd EE (−) 7 ÷ 8 2nd EE 6 ENTER. We have 9×10^{-14}, which checks.

```
7.2E-7/8E6
                9E-14
```

Chapter 6
Polynomials and Factoring

CHECKING FACTORIZATIONS OF POLYNOMIALS

Tables of values are used to check some of the factorizations in **Sections 6.1, 6.2, and 6.3**. The procedure for setting up tables like these is described on page 11 of this manual.

A graph drawn using different GraphStyles is used to check **Example 4** in **Section 6.2**. The procedure for doing this is found on page 21 of this manual.

A horizontal split screen showing a graph and a table is used to check **Example 7** in **Section 6.4**. The procedure for setting up a horizontal split screen is described on pages 22 and 23 of this manual.

SOLVING QUADRATIC EQUATIONS

In **Section 6.6, page 420**, the ZERO option from the CALC menu is used to approximate the solutions of a quadratic equation. The procedure for using this method to solve an equation is found on page 16 of this manual.

Chapter 7
Rational Expressions and Equations

GRAPHING DATA

We use the Edit option from the STAT EDIT menu to enter data which can then be graphed.

Section 7.8, Example 6(a) The f-stop is a setting on a camera that indicates how much light will reach the exposed film. An f-stop is related to the size of the opening of the lens, or aperture. Some f-stops and apertures for a 50 mm lens are shown in the following table. Graph the data and determine whether the data indicate direct variation or inverse variation.

f-stop	Size of aperture, in mm
1.4	35.7
4	12.5
8	6.25
11	4.545

We will enter the data as ordered pairs on the STAT list editor screen. The f-stop will be the first coordinate of each pair and the corresponding size of aperture will be the second coordinate. To clear any existing lists press [STAT] 4 [2nd] [L_1] [,] [2nd] [L_2] [,] [2nd] [L_3] [,] [2nd] [L_4] [,] [2nd] [L_5] [,] [2nd] [L_6] [ENTER]. (L_1 through L_6 are the second operations associated with the numeric keys 1 through 6.) The lists can also be cleared by first accessing the STAT list editor screen by pressing [STAT] [ENTER] or [STAT] 1. These keystrokes display the STAT EDIT menu and then select the Edit option from that menu. Then, for each list that contains entries, use the arrow keys to move the cursor to highlight the name of the list at the top of the column and press [CLEAR] [▽] or [CLEAR] [ENTER].

Once the lists are cleared, we can enter the new data. We will enter the first coordinates in L_1 and the second coordinates in L_2. With the STAT list editor screen displayed as described above, position the cursor at the top of column L_1, below the L_1 heading. To enter 1.4 press 1 [.] 4 [ENTER]. Continue entering the first coordinates 4, 8, and 11, each followed by [ENTER]. The entries can be followed by [▽] rather than [ENTER] if desired. Press [▷] to move to the top of column L_2. Enter the second coordinates 35.7, 12.5, 6.25, and 4.545 in succession, each followed by [ENTER] or [▽]. Note that the coordinates of each point must be in the same position in both lists.

```
L1      L2      L3     2
1.4     35.7    ------
4       12.5
8       6.25
11      4.545
------
L2(5) =
```

Now press [Y =] to go to the equation-editor screen and clear any equations that are currently entered. (See page 10 of this manual.) If you wish, instead of clearing an equation, you can deselect it. (See page 22 of this manual.)

In order to graph the data points we will turn on and define a Stat Plot. To do this, first press [2nd] [STATPLOT] to go to the

STAT PLOTS screen. Press ENTER to select Plot 1 and then position the cursor over On and press ENTER to turn on Plot 1. Next select the scatter diagram for Type, L_1 for Xlist, L_2 for Ylist, and the box for the Mark as shown below. To select Type and Mark, position the cursor over the desired selection and press ENTER. Use the L_1 and L_2 keys (associated with the 1 and 2 numeric keys) to select Xlist and Ylist.

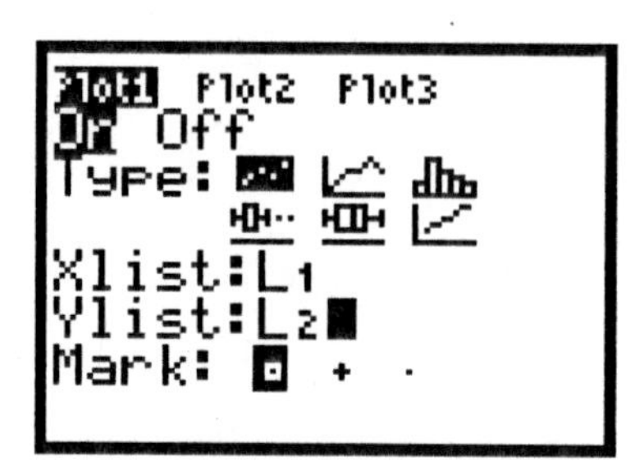

To select the dimensions of the viewing window notice that the f-stops in the table range from 1.4 to 11 and the aperture size ranges from 4.545 to 35.7. We want to select dimensions that will include all of these values. One good choice is [0, 12, 0, 50], Yscl = 10. Enter these dimensions in the WINDOW screen. Then press GRAPH to graph the data. From the graph we see that the data indicate inverse variation.

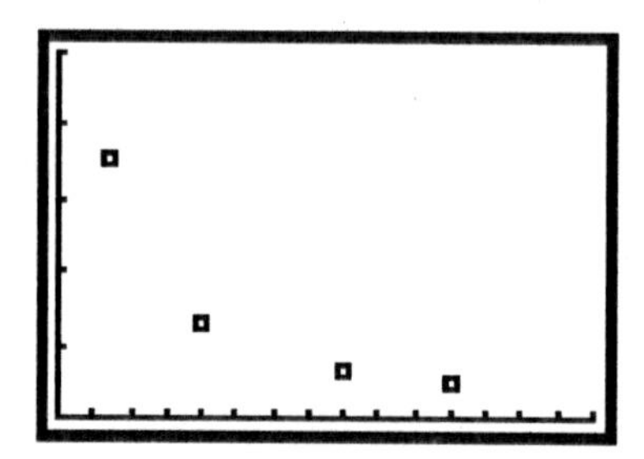

Chapter 8
Radical Expressions and Equations

FINDING SQUARE ROOTS

Section 8.1, Example 5 Use a calculator to approximate $\sqrt{10}$ to three decimal places.

To enter $\sqrt{10}$ on the TI-83, TI-83 Plus, or TI-84 Plus, press [2nd] [$\sqrt{\ }$] 1 0 [)] [ENTER]. ($\sqrt{\ }$ is the second operation associated with the [x^2] key.) We see that $\sqrt{10} \approx 3.162$.

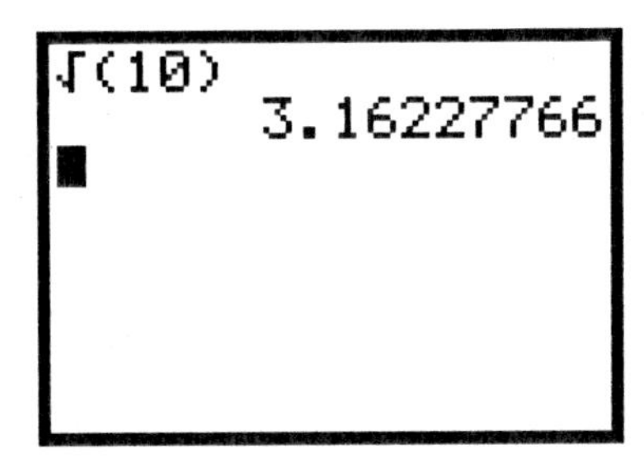

Note that the calculator supplies a left parenthesis along with the radical symbol and we closed the parentheses by adding a right parenthesis after entering the radicand. The calculator considers all the terms that follow the radical symbol to be part of the radicand until a right parenthesis is entered. Thus, although the right parenthesis is not necessary in the example above, we must use a right parenthesis when terms that are not part of the radicand are entered following the radicand. Suppose,for example, that we wanted to compute $\sqrt{9} + 16$. The result is $3 + 16$, or 19. If a right parenthesis is not entered after the 9, the calculator considers the radicand to be $9 + 16$ rather than 9 and computes $\sqrt{9+16}$ or $\sqrt{25}$ as shown below. For this reason it is good to develop the habit of always closing the parentheses in the appropriate place.

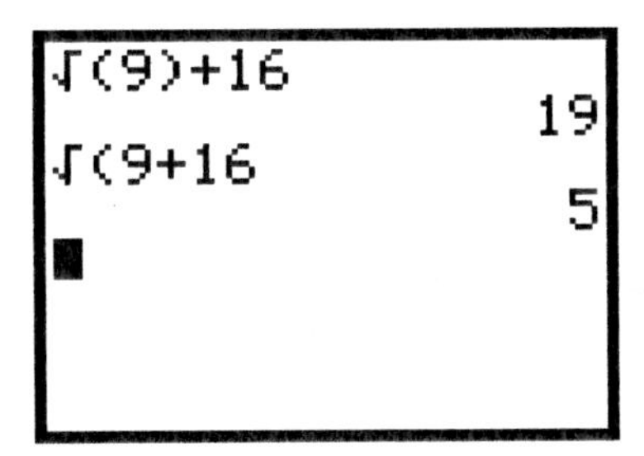

Chapter 9
Quadratic Equations

EVALUATING A FUNCTION

Using function notation to evaluate a function is discussed in the text in **Section 9.7** on **page 612**. For example, to find $f(-1.5)$ when $f(x) = -2x^2 + 3x - 1.2$, first press [Y =] to go to the equation-editor screen. Then mentally replace $f(x)$ with Y_1 and enter $Y_1 = -2x^2 + 3x - 1.2$. Now, to find $f(-1.5)$, or $Y_1(-1.5)$, first press [2nd] [QUIT] to go to the home screen. Then press [VARS] [▷] 1 1 [(] [(−)] 1 [.] 5 [)] [ENTER]. We see that $Y_1(-1.5) = -10.2$, or $f(-1.5) = -10.2$.

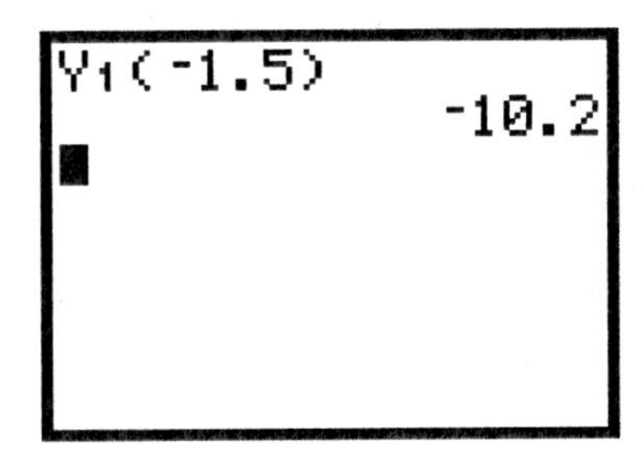

The TI-89
Graphics Calculator

Chapter 1
Introduction to Algebraic Expressions

GETTING STARTED

Press [ON] to turn on the TI-89 graphing calculator. ([ON] is the key at the bottom left-hand corner of the keypad.) The home screen is displayed. You should see a row of boxes at the top of the screen and two horizontal lines with lettering below them at the bottom of the screen. If you do not see anything, try adjusting the display contrast. To do this, first press [◇]. ([◇] is the key in the left column of the keypad with a green diamond inside a green border. All operations associated with the [◇] key are printed on the keyboard in green, the same color as the [◇] key.) Then press [+] to darken the display or [−] to lighten the display. Be sure to use the black [−] key in the right column of the keypad rather than the gray [(−)] key on the bottom row.

One way to turn the calculator off is to press [2nd] [OFF]. (OFF is the second operation associated with the [ON] key. All operations accessed by using the [2nd] key are printed on the keyboard in yellow, the same color as the [2nd] key.) When you turn the TI-89 on again the home screen will be displayed regardless of the screen that was displayed when the calculator was turned off. [2nd] [OFF] cannot be used to turn off the calculator if an error message is displayed. The calculator can also be turned off by pressing [◇] [OFF]. This will work even if an error message is displayed. When the TI-89 is turned on again the display will be exactly as it was when it was turned off. The calculator will turn itself off automatically after several minutes without any activity. When this happens the display will be just as you left it when you turn the calculator on again.

From top to bottom, the home screen consists of the toolbar, the large history area where entries and their corresponding results are displayed, the entry line where expressions or instructions are entered, and the status line which shows the current state of the calculator. These areas will be discussed in more detail as the need arises.

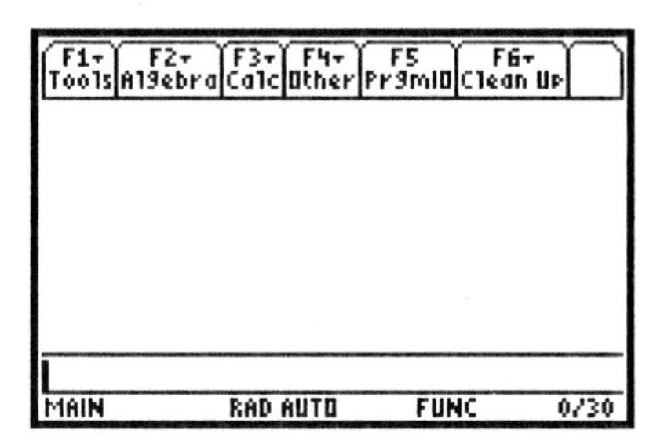

Press [MODE] to display the MODE settings. Modes that are not currently valid, due to the existing choices of settings, are dimmed. Initially you should select the settings shown below.

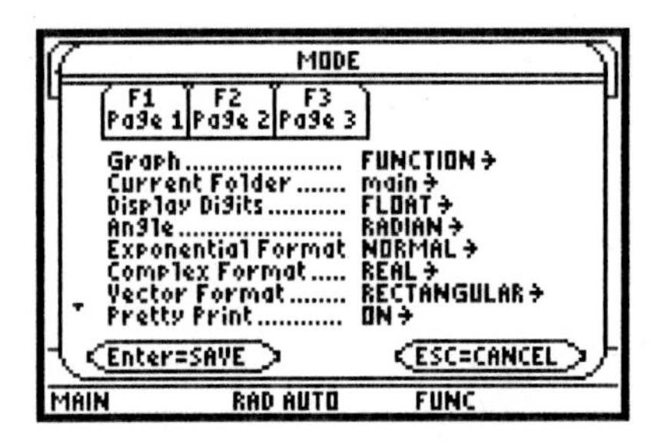

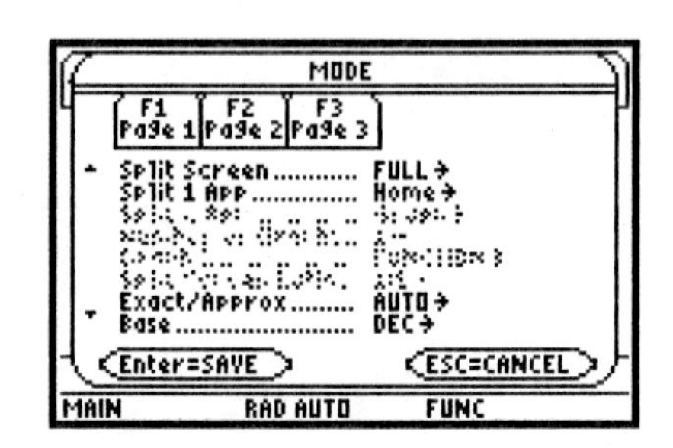

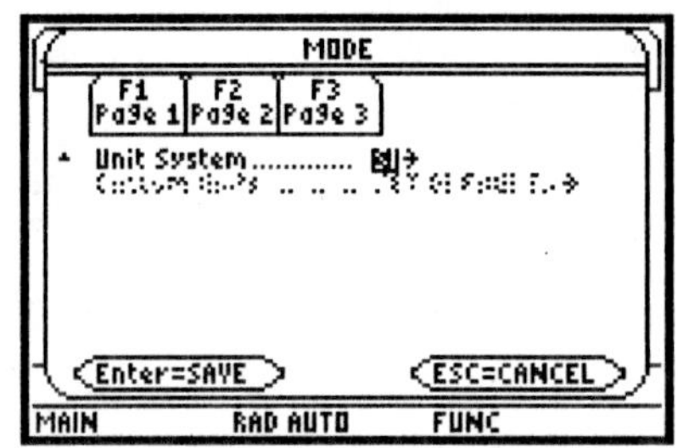

To change a setting on the Mode screen use ▽ or △ to move the cursor to the line of that setting. Then use ▷ to display the options. Press the number of the desired option followed by ENTER. Press HOME or 2nd QUIT to leave the MODE screen and return to the home screen. (QUIT is the second operation associated with the ESC key.) Note that the cursor skips dimmed settings as you move through the options.

It will be helpful to read the Introduction to the Graphing Calculator on pages 4 and 5 of the textbook as well as Chapter 1: Getting Started and Chapter 2: Operating the TI-89 in the TI-89 Guidebook before proceeding.

EVALUATING EXPRESSIONS

To evaluate expressions we substitute values for the variables.

Section 1.1, Example 4 Use a graphing calculator to evaluate $3xy + x$ for $x = 65$ and $y = 92$.

You might want to clear any previously entered computations from the history area of the home screen first. To do this, access Tools from the tool bar at the top of the screen by pressing F1, the blue key at the top left-hand corner of the keypad. Then select item 8, Clear Home, from this menu by pressing 8.

Then enter the expression in the calculator, replacing x with 65 and y with 92. Press 3 × 6 5 × 9 2 + 6 5 ENTER. The calculator returns the value of the expression, 18,005.

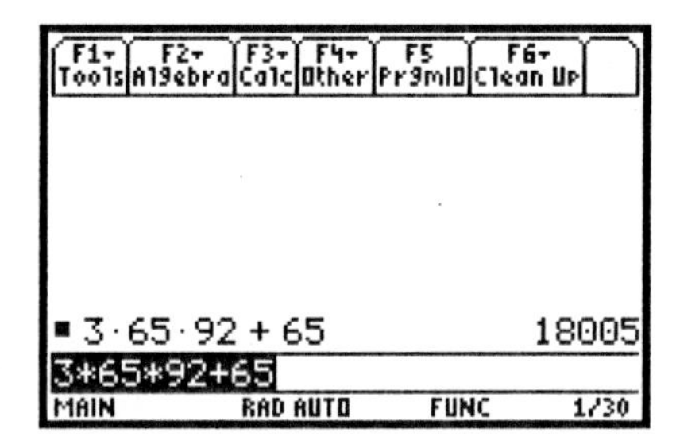

You can edit your entry if necessary. After ENTER is pressed to evaluate an expression, the TI-89 leaves the expression on the entry line and highlights it. To edit the expression you must first remove the highlight to avoid the possibility of accidently typing over the entire expression. To do this, press ◁ or ▷ to move the cursor (a blinking vertical line) toward the side of the expression to be edited. If, for instance, in the expression above you pressed 8 instead of 9, first press ◁ to move the cursor toward the 8. Now, to type a 9 over the 8, first select overtype mode by pressing 2nd INS. (INS is the second operation associated with the ← key.) Now the cursor becomes a dark, blinking rectangle rather than a vertical line. Use ▷ to position the cursor over the 8 and then press 9 to write a 9 over the 8. To leave overtype mode press 2nd INS again. The calculator is now in the insert mode, indicated by a vertical cursor, and will remain in that mode until overtype mode is once again selected.

If you forgot to type the 2, move the insert cursor to the left of the plus sign and press 2 to insert the parenthesis before the plus sign. You can continue to insert symbols immediately after the first insertion. If you typed 31 instead of 3, move the cursor to the left of 1 and press ←. This will delete the 1. Instead of using overtype mode to overtype a character as described above, we can use ← to delete the character and then, in insert mode, insert a new character.

If you accidently press △ instead of ◁ or ▷ while editing an expression, the cursor will move up into the history area of the screen. Press ESC to return immediately to the entry line. The ▽ key can also be used to return to the entry line. It must be pressed the same number of times the △ key was pressed accidently.

If you notice that an entry needs to be edited before you press ENTER to perform the computation, the editing can be done as described above without the necessity of first removing the highlight from the entry.

The keystrokes 2nd ENTRY can be used repeatedly to recall entries preceding the last one. (ENTRY is the second function associated with the ENTER key.) Pressing 2nd ENTRY twice, for example, will recall the next to last entry. Using these keystrokes a third time recalls the third to last entry and so on. The number of entries that can be recalled depends on the amount of storage they occupy in the calculator's memory.

Previous entries and results of computations can also be copied to the entry line by first using the △ key to move through the history area until the desired entry or result is highlighted. Then press ENTER to copy it to the entry line.

USING A MENU

In the previous example we used the Tools menu to clear the home screen. In general, a menu is a list of options that appear when a key is pressed. For example, we pressed F1 to display the Tools menu. We can select an item from a menu by using ▽ to highlight it and then pressing ENTER or by simply pressing the number of the item. If an item is identified by a letter rather than a number, press the purple alpha key followed by the letter of the item. The letters are printed in purple above the keys on the keypad. The down-arrow beside item 8 in the menu above indicates that there are additional items in the menu. Use ▽ to scroll down to them.

FRACTION NOTATION

Section 1.3, Example 12 Use a graphing calculator to find fraction notation for $\frac{2}{15} + \frac{7}{12}$.

Go to the entry line of the home screen and press 2 [÷] 1 5 [+] 7 [÷] 1 2 [ENTER].

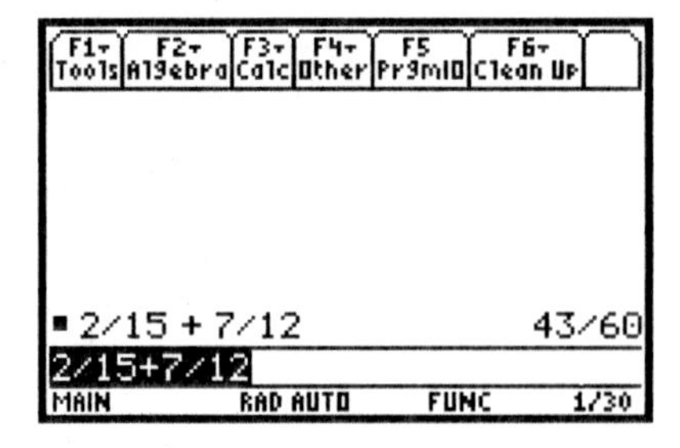

Note that the result is expressed in fraction notation. This will occur when Exact is selected as the Exact/Approx mode on the Mode screen. It will also occur when the Auto setting is selected and there is no decimal point in the entry.

To see the result expressed in decimal notation, we can enter the expression as above and then press [◇] before pressing [ENTER]. Note the black ◇ at the bottom of the screen on the left below and the decimal result on the right.

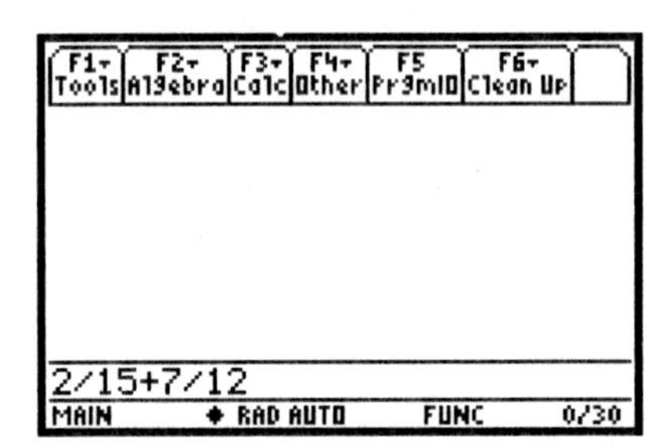

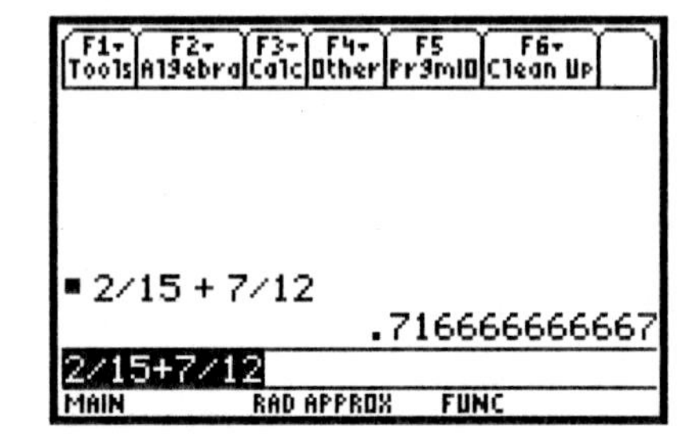

We will also see the result expressed in decimal notation if Approximate is selected as the Exact/Approx mode before entering the expression.

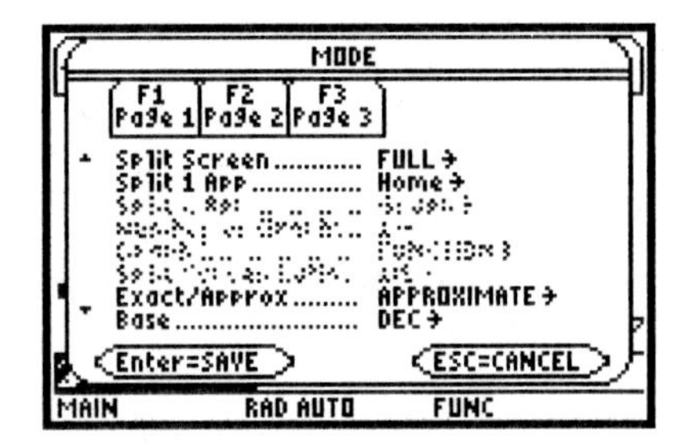

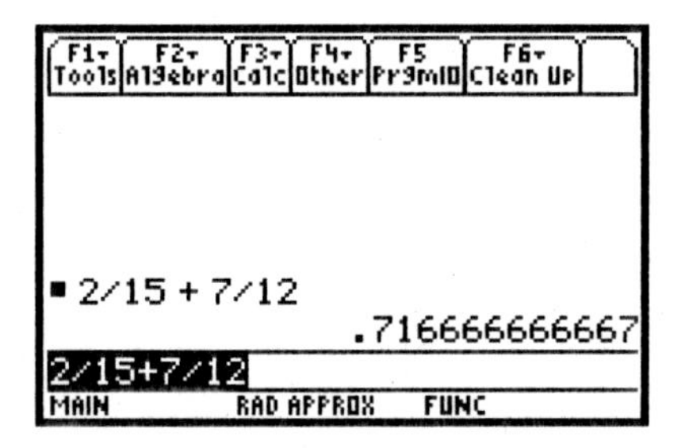

In addition, with Auto mode selected, the result will be expressed in decimal notation if we include a decimal point with one of the integers in the expressions. If we enter $\frac{2}{15} + \frac{7}{12.}$ in Auto mode, for example, the result appears in decimal notation.

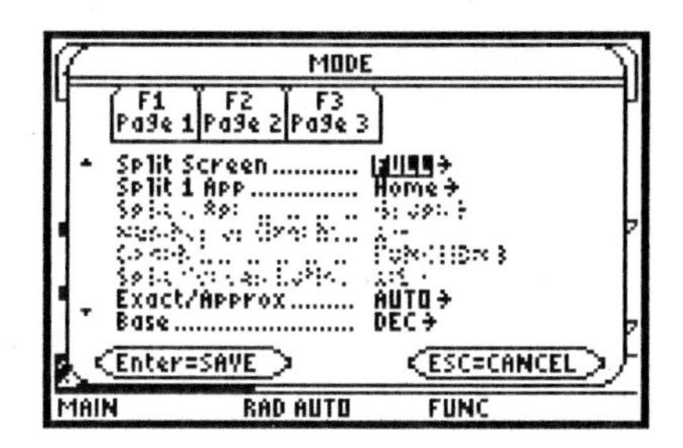

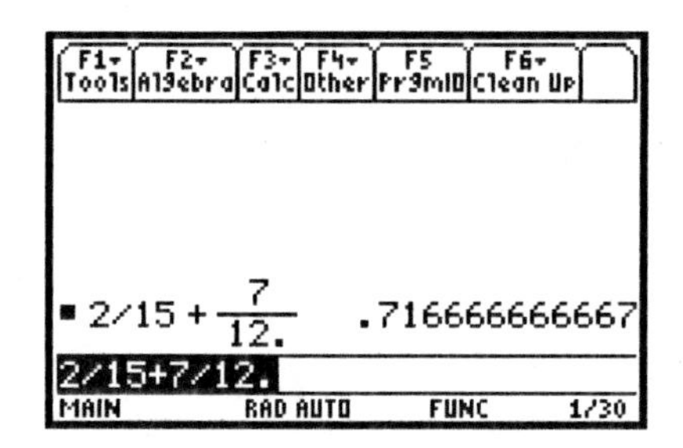

To convert a fractional result to decimal form, use $\boxed{\triangle}$ to move into the history area and highlight the result. Press $\boxed{\text{ENTER}}$ to copy the result to the entry line. Then press $\boxed{\diamond}$ $\boxed{\text{ENTER}}$ to see the decimal form of the result. This occurs regardless of the Exact/Approx setting. (Note that we ordinarily set the calculator in Auto mode.)

SQUARE ROOTS

Section 1.4, Example 5 Graph the real number $\sqrt{3}$ on a number line.

We can use the calculator to find a decimal approximation for $\sqrt{3}$. If the calculator is in Exact or Auto mode, press $\boxed{\text{2nd}}$ $\boxed{\sqrt{}}$ 3 $\boxed{)}$ $\boxed{\diamond}$ $\boxed{\text{ENTER}}$. If the calculator is in Approximate mode, the $\boxed{\diamond}$ is not necessary. Note that the calculator supplies a left parenthesis along with the radical symbol. We must supply the right parenthesis to complete the expression.

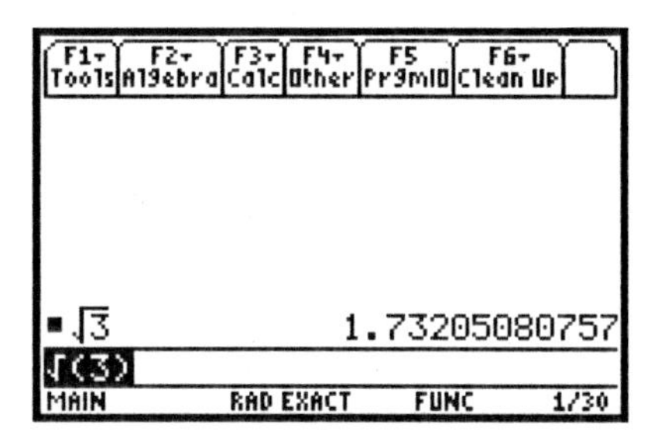

This approximation can now be used to locate $\sqrt{3}$ on the number line. The graph appears on page 33 of the text.

NEGATIVE NUMBERS AND ABSOLUTE VALUE

On the entry line of the TI-89, $|x|$ is written abs(x). Since we selected Pretty Print mode earlier, the traditional notation will appear in the history area.

Section 1.4, Example 8 Find each absolute value: **(a)** $|-3|$; **(b)** $|7.2|$; **(c)** $|0|$.

Absolute value notation is the first item in the graphing calculator's catalog, an alphabetic list of all the commands on the TI-89. When entering $|-3|$, keep in mind that the gray $\boxed{(-)}$ key in the bottom row of the keypad must be used to enter a negative number on the calculator whereas the black $\boxed{-}$ key in the right-hand column of the keypad is used to enter subtraction. Go to the entry line on the home screen and press $\boxed{\text{CATALOG}}$ $\boxed{\text{ENTER}}$ $\boxed{(-)}$ 3 $\boxed{)}$ $\boxed{\text{ENTER}}$. To find $|7.2|$ use the keystrokes above, replacing $\boxed{(-)}$ 3 with 7 $\boxed{.}$ 2. Note that, in order to get the result in decimal form, $\boxed{\diamond}$ must also be pressed before $\boxed{\text{ENTER}}$ if the calculator is set in Exact mode. To find $|0|$ replace $\boxed{(-)}$ 3 with 0. Note that the calculator supplies a left parenthesis along with the absolute value notation. We must supply the right parenthesis to close the absolute-value expression.

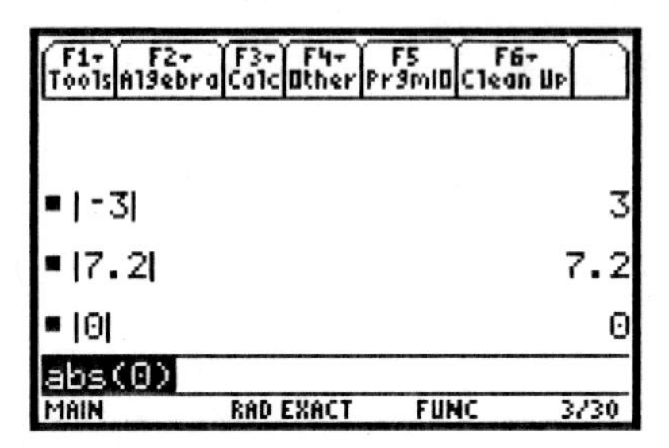

ORDER OF OPERATIONS; EXPONENTS AND GROUPING SYMBOLS

The TI-89 follows the rules for order of operations.

Section 1.8, Example 8 Calculate: $\dfrac{12(9-7)+4\cdot 5}{2^4+3^2}$.

The fraction bar must be replaced with a set of parentheses around the entire numerator and another set of parentheses around the entire denominator when this expression is entered in the calculator. To enter an exponential expression, first enter the base, then use the ∧ key followed by the exponent.

To enter the expression above and express the result in fraction notation, set the calculator in Exact or Auto mode and press (1 2 (9 − 7) + 4 × 5) ÷ (2 ∧ 4 + 3 ∧ 2). Remember to use the black − key for subtraction rather than the gray (−) key, which is used to enter negative numbers.

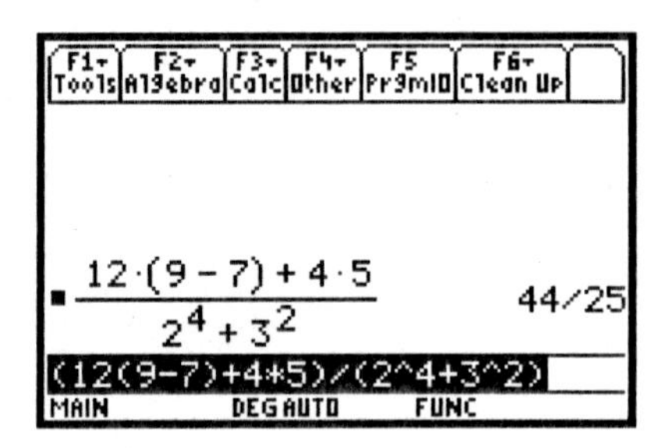

Note that although the ∧ symbol appears in the expression on the entry line, the history area shows the exponent in the traditional format. This happens because we selected Pretty Print mode earlier.

Chapter 2
Equations, Inequalities, and Problem Solving

EDITING ENTRIES

In **Section 2.1** of the text the procedure for editing entries is discussed on page 75. This procedure is described on page 39 of this manual.

EVALUATING EXPRESSIONS

Section 2.1, Example 3 Use a calculator to determine whether each of the following is a solution of $2x - 5 = -7$; $3, -1$.

For a given value of x, if the value of the expression on the left side of the equation, $2x - 5$, is the same as the number on the right side of the equation, -7, then the given value of x is a solution of the equation. We can evaluate the expression on the left side for each value of x directly by substituting that value for x as described on page 38 of this manual. We could also evaluate the expression $2x - 5$ by storing a value of x in the calculator first and then entering the expression in terms of the variable.

To store 3 as x, press 3 [STO▷] [X] [ENTER]. Then enter the expression $2x - 5$ by pressing 2 [X] [−] 5. Finally, press [ENTER] to evaluate the expression. We see that $2x - 5$ is 1 when $x = 3$, so 3 is not a solution of the equation $2x - 5 = -7$. Next store -1 as x by pressing [(−)] 1 [STO▷] [X] [ENTER]. Then press [2nd] [ENTRY] to recall the expression $2x - 5$ to the screen. Now press [ENTER]. We see that $2x - 5$ is -7 for $x = -1$, so -1 is a solution of the equation $2x - 5 = -7$.

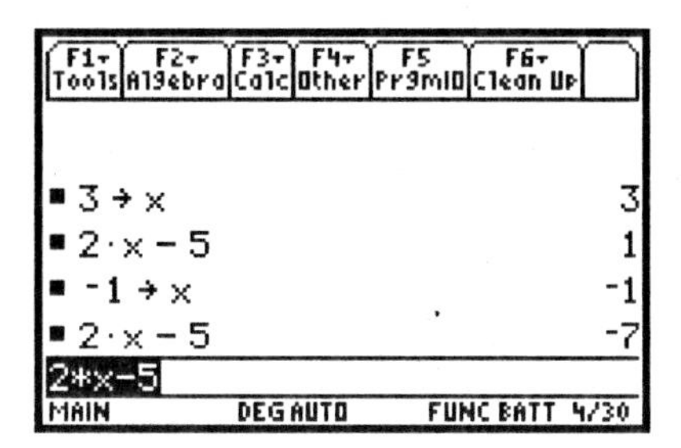

EVALUATING FORMULAS; THE TABLE FEATURE: ASK MODE

Section 2.3, Example 2 Use the formula $B = 30a$, described in Example 1, to determine the minimum furnace output for well-insulated houses containing 800 ft^2, 1500 ft^2, 2400 ft^2, and 3600 ft^2.

First we replace B with y and a with x and enter the formula $y = 30x$ on the equation-editor screen as equation y_1. Press [◇] [Y =] to access this screen. If a plot is currently turned on, it should be turned off, or deselected, now. A check mark beside a plot indicates that it is currently selected. To deselect it, move the cursor to the plot on the equation-editor screen. Then press [F4]. There should now be no check mark beside the plot, indicating that it has been deselected. Do this for each plot that has a check mark beside it. If there is currently an expression displayed for y_1, clear it by positioning the cursor beside "y1 =" and pressing [CLEAR]. Do the same for expressions that appear on all other "y =" lines by using [▽] to move to a line and then pressing [CLEAR]. Then use [△] or [▽] to move the cursor beside "y1 =." Now enter $y_1 = 30x$ on the entry line of the equation-editor screen and paste it beside "y1 =" by pressing 3 0 [X] [ENTER].

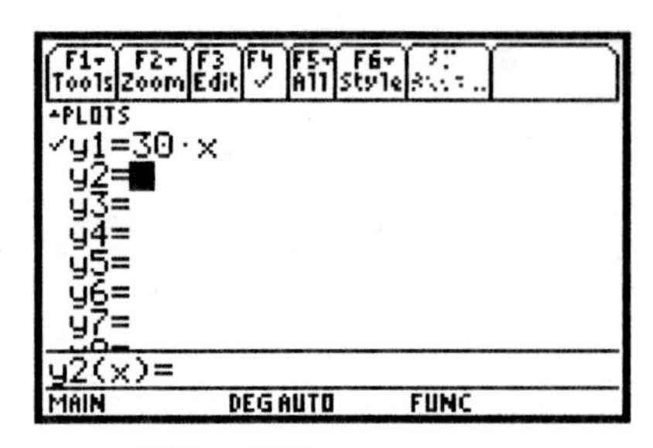

To edit an entry on the equation-editor screen, use △ or ▽ to highlight it and then press ENTER. This copies the entry to the entry line where it can be edited as described on page 39 of this manual.

For an equation entered in the equation-editor screen, a table of x-and y-values can be displayed. We will use a table to evaluate the formula for the given values. Once the equation is entered, press ◇ TblSet to display the Table Setup screen. (TblSet is the green ◇ operation associated with the F4 key.) The Table Setup screen can also be accessed by pressing ◇ TABLE F2. (TABLE is the green ◇ operation associated with the F5 key.) You can choose to supply the x-values yourself or you can set the calculator to supply them. To choose the x-values yourself, move the cursor to the "Independent" line. Then press ▷ 2 ENTER to select Ask mode. In Ask mode the calculator disregards the other settings on the Table Setup screen.

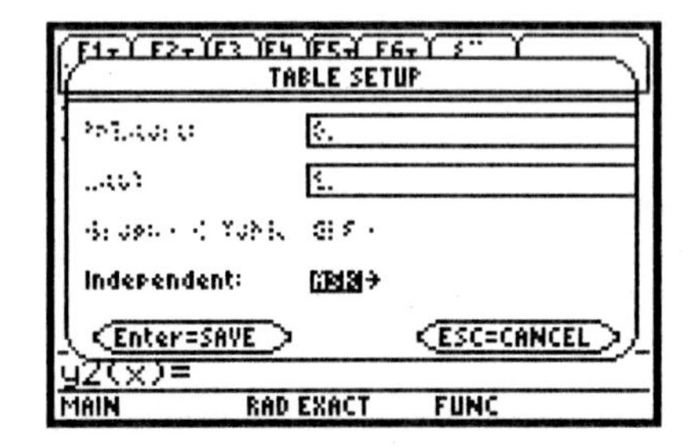

Now press ◇ TABLE to view the table. If you select Ask before a table is displayed for the first time on your calculator, a blank table is displayed. If a table has previously been displayed, the table you now see will continue to show the values in the previous table.

Values for x can be entered in the x-column of the table and the corresponding values for y_1 will be displayed in the $y1$-column. To enter 800, 1500, 2400, and 3600 press 8 0 0 ENTER ▽ 1 5 0 0 ENTER ▽ 2 4 0 0 ENTER ▽ 3 6 0 0 ENTER. Any additional x-values that are displayed are from a table that was previously displayed on the Auto setting. The y-values are displayed in the table in scientific notation rounded to one decimal place. (See page 364 in the text for a discussion of scientific notation.) To see the exact values move the cursor to the entries and read the values displayed at the bottom of the screen.

x	y1
800.	2.4E4
1500.	4.5E4
2400.	7.2E4
3600.	1.1E5

y1(x)=108000.

MAIN DEG AUTO FUNC

We see that $y_1 = 24,000$ when $x = 800$, $y_1 \doteq 45,000$ when $x = 1500$, $y_1 = 72,000$ when $x = 2400$, and $y_1 = 108,000$ when $x = 3600$, so the furnace outputs for 800 ft^2, 1500 ft^2, 2400 ft^2, and 3600 ft^2 are 24,000 Btu's, 45,000 Btu's, 72,000 Btu's, and 108,000 Btu's, respectively.

THE TABLE FEATURE: AUTO MODE

Section 2.4, Example 8 Village Stationers wants the customer service team to be able to look up the cost c of merchandise being returned when only the total amount paid T (including tax) is shown on the receipts. Use the formula $c = T/1.05$, developed in Example 7, to create a table of values showing cost given a total amount paid. Assume that the least possible sale is $0.21.

We will use the graphing calculator to create the table of values. To enter the formula, we first replace c with y and T with x. Then we enter the formula on the equation-editor screen as $y = x/1.05$. Be sure that the Plots are turned off and that any previous entries are cleared. (See page 43 of this manual for the procedure for turning off the Plots and clearing equations.)

F1 Tools | F2 Zoom | F3 Edit | F4 ✓ | F5 All | F6 Style
PLOTS
✓y1 = x / 1.05
y2=
y3=
y4=
y5=
y6=
y2(x)=
MAIN RAD EXACT FUNC

We will create a table in which the calculator supplies the x-values beginning with a value we specify and continuing by adding a value we specify to the preceding value for x. We will begin with an x-value of 0.21 (corresponding to $0.21) and choose successive increases of 0.01 (corresponding to $0.01).

To do this, first press [◇] [TblSet] or [◇] [TABLE] [F2] to access the TABLE SETUP window. If "Independent" is set to "Auto" on the Table Setup screen, the calculator will supply values for x, beginning with the value specified as tblStart and continuing by adding the value of Δtbl to the preceding value for x. If the table was previously set to Ask, the blinking cursor will be positioned over ASK. Change this setting to AUTO by pressing [▷] 1 [ENTER]. Now the Table Setup screen must once again be accessed so that we can set tblStart and Δtbl. Enter a minimum x-value of 0.21, an increment of 0.01, and a Graph < – > Table setting of OFF by first positioning the cursor beside tblStart and then pressing [.] 2 1 [▽] [.] 0 1 [▽] [▷] 1 [ENTER]. Press [◇] [TABLE] to see the table.

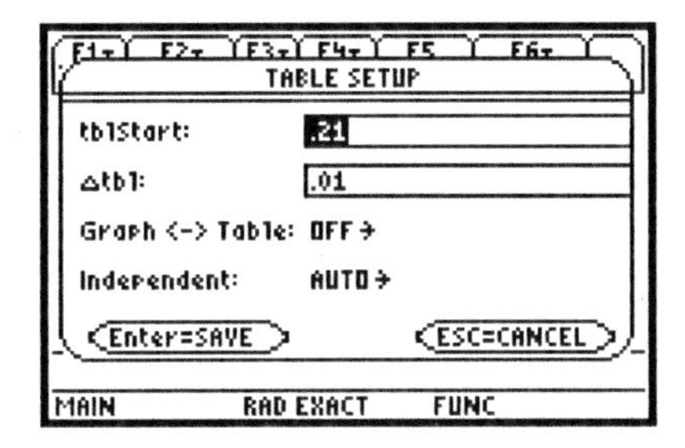

x	y1
.21	.2
.22	.20952
.23	.21905
.24	.22857
.25	.2381

x=.21
MAIN RAD EXACT FUNC

We can change the number of decimal places that the calculator will display to 2 so that the costs are rounded to the nearest cent. To do this, press [MODE] to access the MODE screen. Then press [▽] [▽] to move the cursor to the third line, Display Digits, and press the [▷] key to display the options. Then use [▽] or [△] to highlight FLOAT 2. Finally, press [ENTER] [ENTER] to select and save two decimal places on the MODE screen.

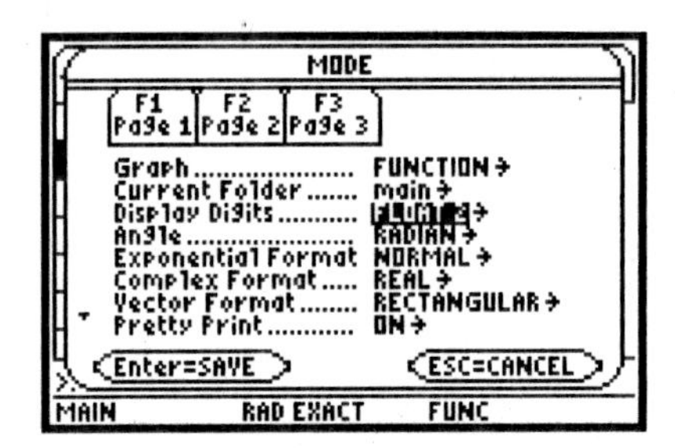

x	y1
.21	.2
.22	.21
.23	.22
.24	.23
.25	.24

x=.21

MAIN RAD EXACT FUNC

Use the ▽ and △ keys to scroll through the table.

Before proceeding, return to the MODE screen and reselect "Float." This allows the number of decimal places to vary, or float, according to the computation being performed.

In **Section 2.5, Example 2**, two equations are entered on the equation-editor screen. The first equation, $y_1 = x + 1$, can be entered as described on page 43 of this manual. Entering y_1 moves the cursor beside "y2 =." Now enter the right-hand side of the second equation, $x + (x + 1)$. Note that the equation-editor screen will not display the parentheses even if they are keyed in on the entry line.

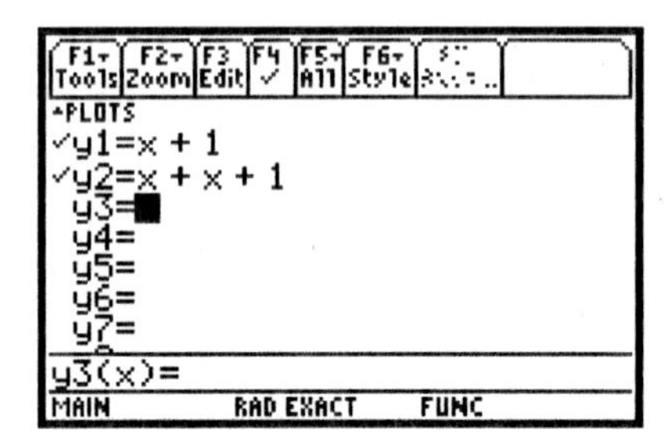

Chapter 3
Graphs and Linear Equations

If you selected 2 for the number of decimal places in Section 2.4, Example 8, and have not yet returned to the MODE screen to reselect "Float," do so now. In Float mode the number of decimal places varies, or floats, according to the computation being performed.

SETTING THE VIEWING WINDOW

The viewing window is the portion of the coordinate plane that appears on the calculator's screen. It is defined by the minimum and maximum values of x and y: xmin, xmax, ymin, and ymax. The notation [xmin, xmax, ymin, ymax] is used in the text to represent these window settings or dimensions. For example, $[-12,\ 12,\ -8,\ 8]$ denotes a window that displays the portion of the x-axis from -12 to 12 and the portion of the y-axis from -8 to 8. In addition, the distance between tick marks on the axes is defined by the settings xscl and yscl. In this manual xscl and yscl will be assumed to be 1 unless noted otherwise. The setting xres sets the pixel resolution. We usually select xres = 2. The window corresponding to the settings $[-20,\ 30,\ -12,\ 20]$, xscl = 5, yscl = 2, xres = 2, is shown below.

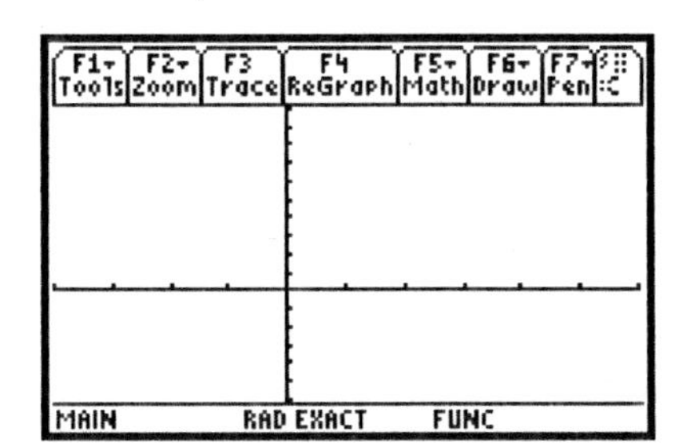

Press [◇] [WINDOW] to display the current window settings on your calculator. (WINDOW is the ◇ operation associated with the [F2] key on the top row of the keypad.) The standard settings are shown below.

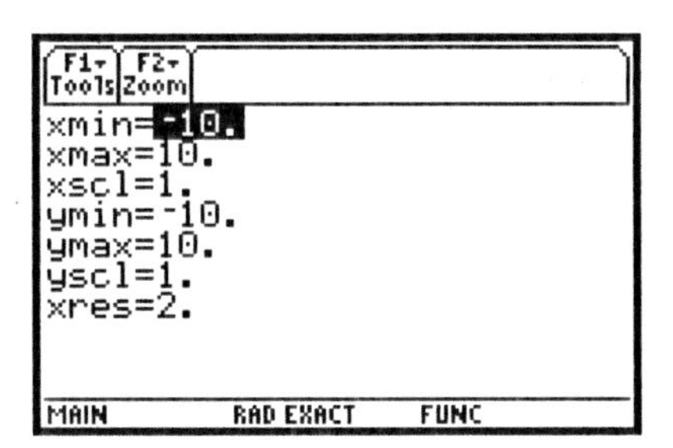

Section 3.1, Example 5 Set up a $[-100, 100, -5, 5]$ viewing window on a graphing calculator, choosing appropriate scales for the axes.

To change a setting, position the cursor beside the setting you wish to change and enter the new value. We will enter the given settings and let xscl = 10 and yscl =1. The choice of scales for the axes may vary. We select scaling that will allow some space between tick marks. If a scale that is too small is chosen, the tick marks will blend and blur. To change the settings to $[-100,\ 100,\ -5,\ 5]$, xscl = 10, yscl =1, on the Window screen, start with the setting beside xmin = highlighted and press [(−)]

1 0 0 ENTER 1 0 0 ENTER 1 0 ENTER (−) 5 ENTER 5 ENTER 1 ENTER. The ▽ key may be used instead of ENTER after typing each window setting. To see the window, press ◇ GRAPH. (GRAPH is the ◇ operation associated with the F3 key on the top row of the keypad.)

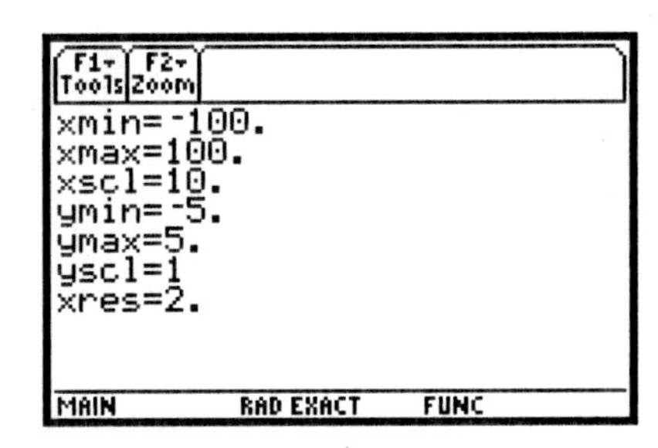

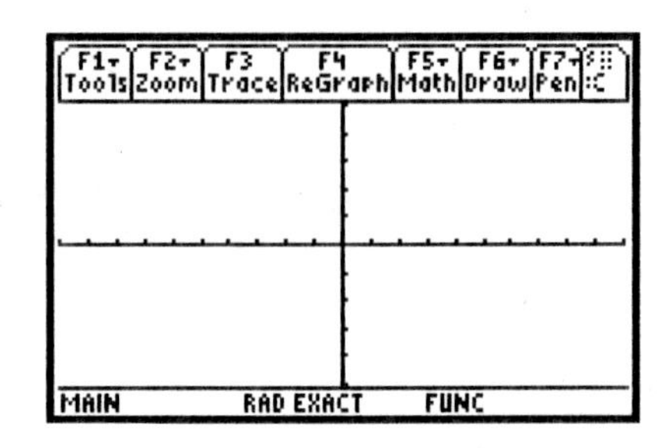

QUICK TIP: To return quickly to the standard window setting [−10, 10, −10, 10], xscl = 1, yscl = 1, when either the Window screen or the Graph screen is displayed, press F2 to access the ZOOM menu and then press 6 to select item 6, ZoomStd (Zoom Standard).

GRAPHING EQUATIONS

After entering an equation and setting a viewing window, you can view the graph of an equation.

Section 3.2, Example 4 Graph $y = 2x$ using a graphing calculator.

Equations are entered on the equation-editor screen. Press ◇ Y = to access this screen. If Plot 1 was used in the example above, it should be turned off, or deselected, now. A check mark beside Plot 1 indicates that it is currently selected. To deselect it, move the cursor to Plot 1 on the equation-editor screen. Then press F4. There should now be no check mark beside Plot 1, indicating that it has been deselected. If there is currently an expression displayed for y_1, clear it as described above. Do the same for expressions that appear on all other "y =" lines by using ▽ to move to a line and then pressing CLEAR. Then use △ or ▽ to move the cursor beside "y1 =." Now enter $y_1 = 2x$ on the entry line of the equation-editor screen and paste it beside "y1 =" by pressing 2 X ENTER.

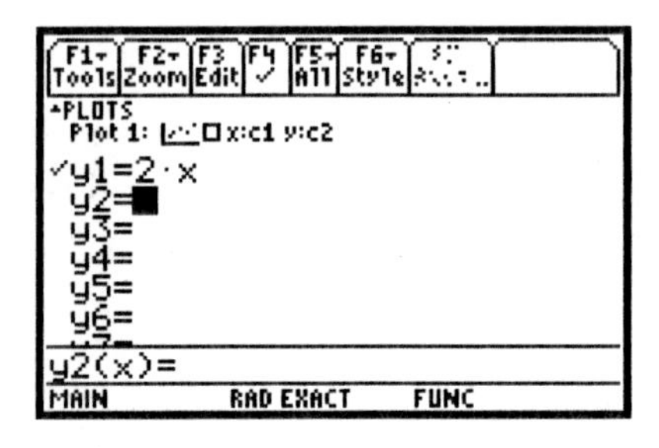

The standard [−10, 10, −10, 10] window is a good choice for this graph. Either enter these dimensions in the WINDOW screen and then press ◇ GRAPH to see the graph or simply press F2 6 to select the standard window and see the graph.

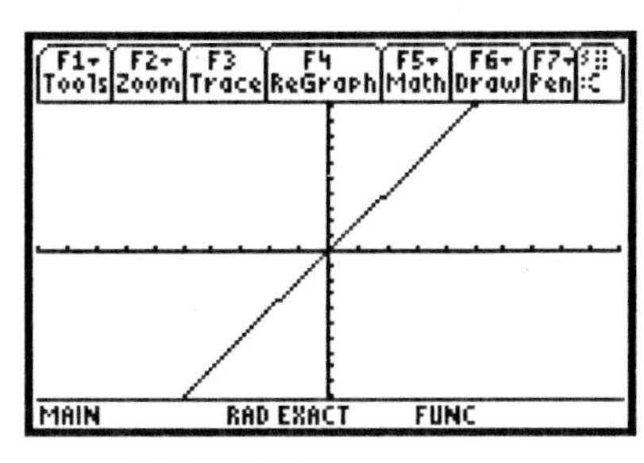

To edit an entry on the equation-editor screen, use △ or ▽ to highlight it and then press ENTER. This copies the entry to the entry line where it can be edited as described on page 39 of this manual.

SOLVING EQUATIONS GRAPHICALLY: THE INTERSECTION METHOD

We can use the Intersection feature from the Math menu on the Graph screen to solve equations.

Section 3.3, Example 2 Solve using a graphing calculator: $-\frac{3}{4}x + 6 = 2x - 1$.

On the equation-editor screen, clear any existing entries and then enter $y_1 = -\frac{3}{4}x + 6$ and $y_2 = 2x - 1$. Press F2 6 to graph these equations in the standard viewing window. The solution of the equation $-\frac{3}{4}x + 6 = 2x - 1$ is the first coordinate of the point of intersection of these graphs. To use the Intersection feature to find this point, first press F5 5 to select Intersection from the Math menu on the Graph screen. The query "1st curve?" appears at the bottom of the screen. The blinking cursor is positioned on the graph of y_1. This is indicated by the 1 in the upper right-hand corner of the screen. Press ENTER to indicate that this is the first curve involved in the intersection. Next the query "2nd curve?" appears at the bottom of the screen. The blinking cursor is now positioned on the graph of y_2 and the notation 2 should appear in the top right-hand corner of the screen. Press ENTER to indicate that this is the second curve. We identify the curves for the calculator since we could have more than two graphs on the screen at once. After we identify the second curve, the query "Lower bound?" appears at the bottom of the screen. Use the right and left arrow keys to move the blinking cursor to a point to the left of the point of intersection of the lines or type an x-value less than the x-coordinate of the point of intersection. Then press ENTER. Next the query "Upper bound?" appears. We give a lower and an upper bound since some pairs of curves have more than one point of intersection. Move the cursor to a point to the right of the point of intersection or type an x-value greater than the x-value of the point of intersection and press ENTER. Now the coordinates of the point of intersection appear at the bottom of the screen.

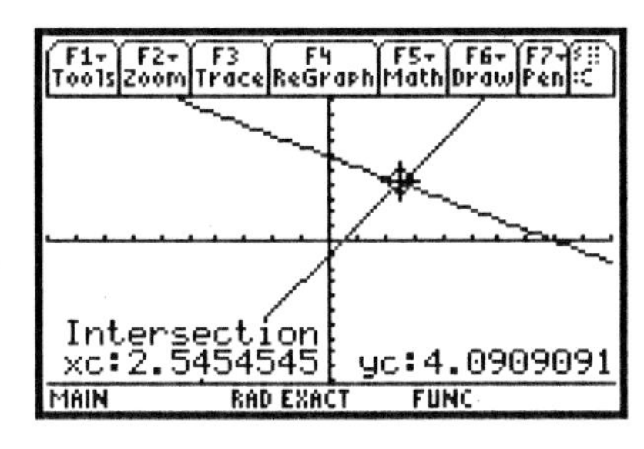

We see that $x = 2.5454545$, so the solution of the equation is 2.5454545.

We can check the solution by evaluating both sides of the equation $-\frac{3}{4}x + 6 = 2x - 1$ for this value of x. The first coordinate of the point of intersection has automatically been stored as xc in the calculator, so we evaluate y1 and y2 for this value of x. First press HOME or 2nd QUIT to go to the home screen. Then to evaluate y1 press Y 1 (X alpha C) ENTER. To evaluate y_2 press Y 2 (X alpha C) ENTER. We see that y1 and y2 have the same value when $x = 2.5454545$, so the solution checks. If your calculator is set in Exact mode also press ◇ before pressing ENTER in order to see the result in decimal notation.

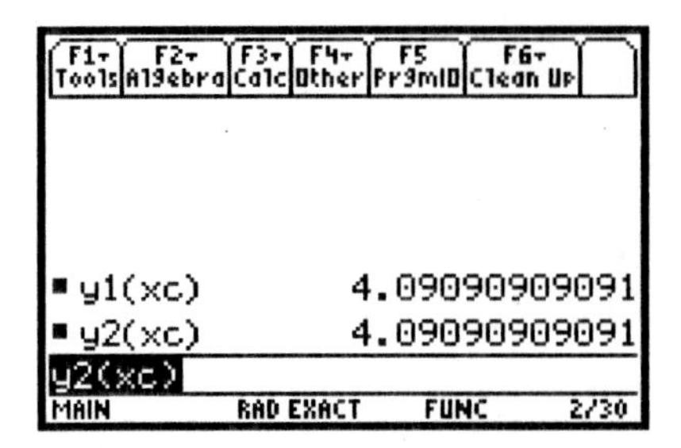

Note that although the procedure above verifies that 2.5454545 is the solution, it is actually an approximation of the solution. To find the exact solution we can solve the equation algebraically.

SOLVING EQUATIONS GRAPHICALLY: THE ZERO METHOD

As an alternative to the Intersection method, we can use the Zero feature from the Math menu on the Graph screen to solve equations.

Section 3.3, Example 7 Solve $3 - 8x = 5 - 7x$ using the zero method.

First we get zero on one side of the equation:

$$3 - 8x = 5 - 7x$$
$$-2 - 8x = -7x$$
$$-2 - x = 0.$$

Then we graph $y = -2 - x$. We will use the standard window.

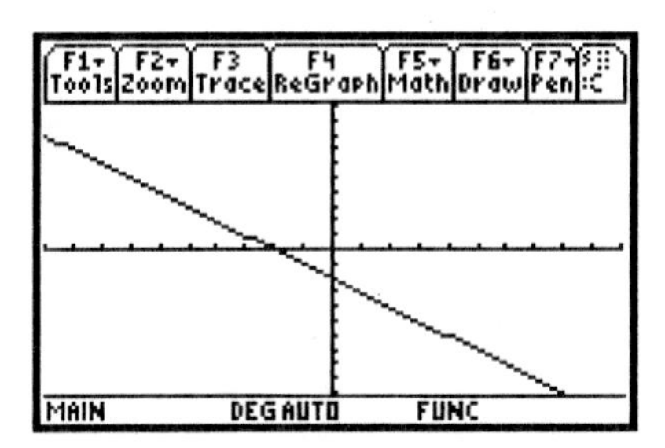

The first coordinate of the x-intercept of the graph is the zero of $y = -2 - x$ and thus is the solution of the original equation. Press 2nd F5 2 to select Zero from the Math menu on the Graph screen. We are prompted to select a lower bound. This means that we must choose an x-value that is to the left of the first coordinate of the x-intercept. This can be done by using the left and right arrow keys to move the cursor to a point on the graph that is to the left of the intercept or by keying in an x-value that is less than the first coordinate of the intercept.

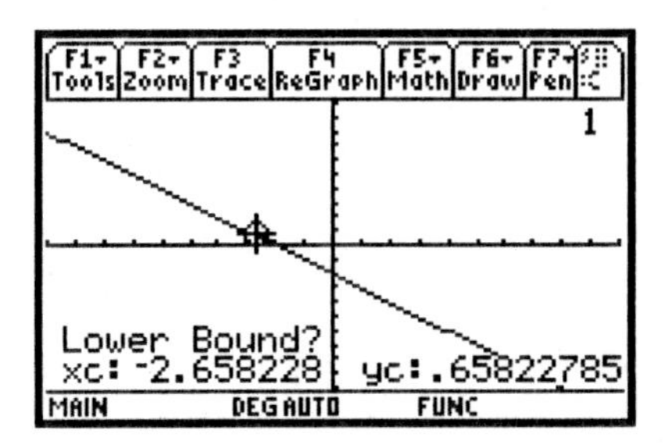

Once this is done, press ENTER. Now we are prompted to select an upper bound that is to the right of the x-intercept. Again we can use the arrow keys or key in a value.

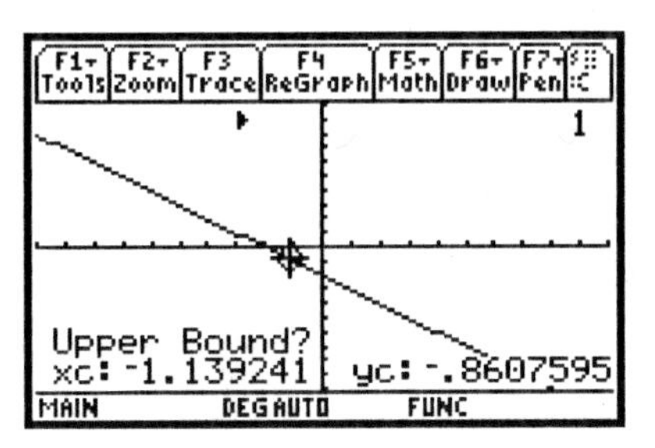

Press ENTER again. We see that $y = 0$ when $x = -2$, so -2 is the zero of $y = -2 - x$ and is thus the solution of the original equation, $3 - 8x = 5 - 7x$.

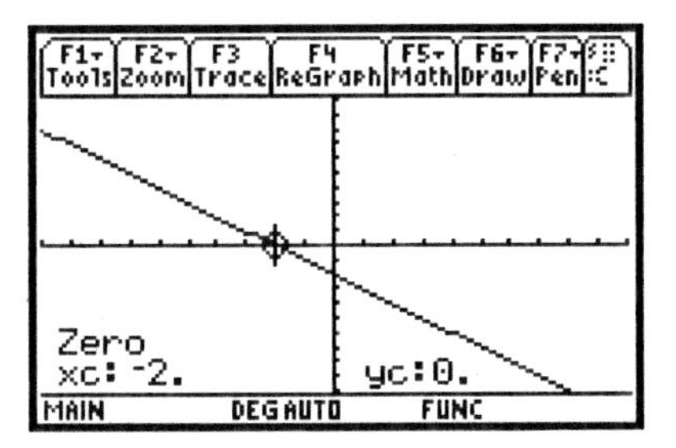

SQUARING THE VIEWING WINDOW

Section 3.7, Example 9 Determine whether the lines given by the equations $3x - y = 7$ and $x + 3y = 1$ are perpendicular, and check by graphing.

In the text each equation is solved for y in order to determine the slopes of the lines. We have $y = 3x - 7$ and $y = -\frac{1}{3}x + \frac{1}{3}$. Since $3\left(-\frac{1}{3}\right) = -1$, we know that the lines are perpendicular. To check this, we graph $\text{y1} = 3x - 7$ and $\text{y2} = -\frac{1}{3}x + \frac{1}{3}$. The graphs are shown on the right below in the standard viewing window.

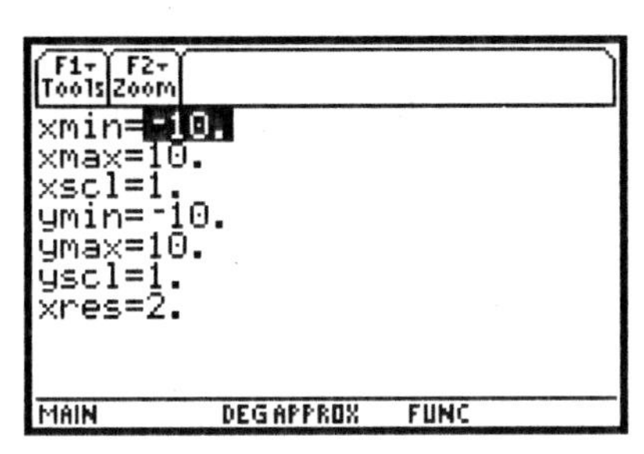

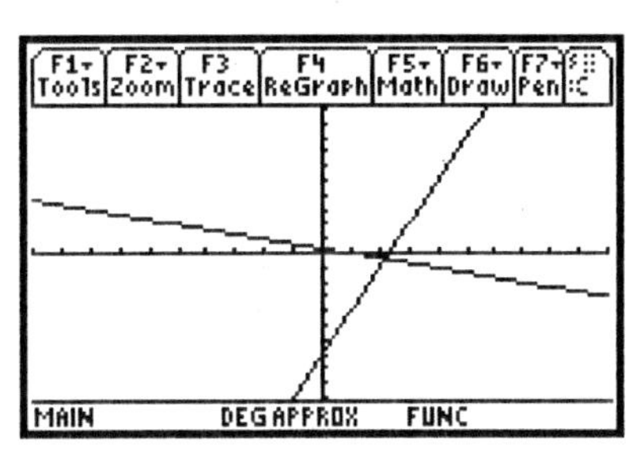

Note that the graphs do not appear to be perpendicular. This is due to the fact that, in the standard window, the distance between tick marks on the y-axis is about 1/2 the distance between tick marks on the x-axis. It is often desirable to choose window dimensions for which these distances are the same, creating a "square" window. On the TI-89 any window in which the ratio of the length of the y-axis to the length of the x-axis is 1/2 will produce this effect. This can be accomplished by selecting dimensions for which $\text{ymax} - \text{ymin} = \frac{1}{2}(\text{xmax} - \text{xmin})$. For example, the windows $[-12, 12, -6, 6]$ and $[-6, 6, -3, 3]$ are square.

When we change the window dimensions to $[-12, 12, -6, 6]$ and press ◇ GRAPH, the graphs now appear to be perpendicular as shown on the right below. From the equation-editor, Window, or Graph screen, we could also press F2 5 to select ZoomSqr. When this is done, the calculator will select a square window.

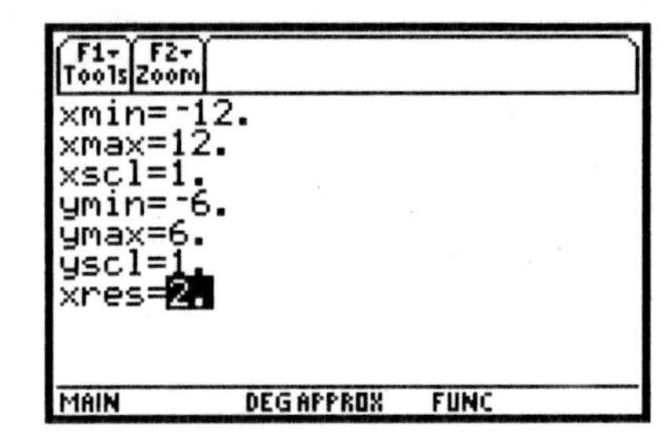
F1▾ Tools
F2▾ Zoom
xmin=-12.
xmax=12.
xscl=1.
ymin=-6.
ymax=6.
yscl=1.
xres=2.
MAIN DEG APPROX FUNC

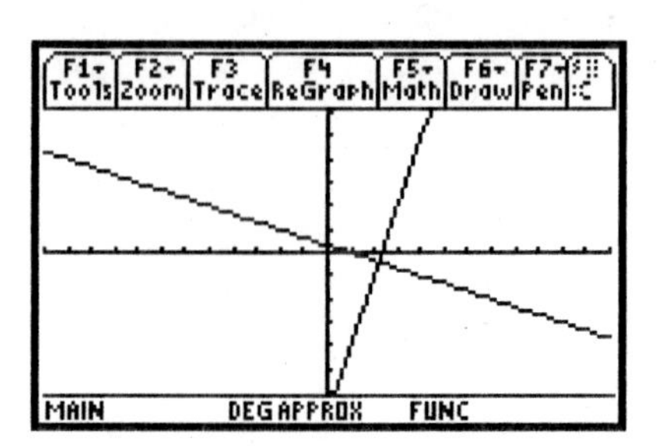
F1▾ Tools
F2▾ Zoom
F3 Trace
F4 ReGraph
F5▾ Math
F6▾ Draw
F7▾ Pen
MAIN DEG APPROX FUNC

Chapter 4
Systems of Equations and Problem Solving

SOLVING SYSTEMS OF EQUATIONS BY GRAPHING

The intersection feature can be used to solve a system of two equations in two variables. The procedure for finding the point of intersection of the graphs of two equations is described on page 49 of this manual.

GRAPHING INEQUALITIES IN TWO VARIABLES

The solution set of an inequality in two variables can be graphed on the TI-89.

Section 4.5, Example 5 Graph $2x + 3y \leq 6$ using a graphing calculator.

Solving for y, we have $y \leq -\frac{2}{3}x + 2$. Then the boundary line is $y = -\frac{2}{3}x + 2$. We will enter this as $y1$. Press [◇] [Y =] to go to the equation-editor screen. If there is currently an entry for $y1$, clear it. Also clear any other equations that are entered. Now enter $y_1 = -\frac{2}{3}x + 2$. Since the inequality is in the form $y \leq mx + b$, we shade the half-plane below the graph of y_1. Eight graph styles can be selected on the TI-89. The "Below" style can be used to shade the region below the boundary line. To do this, use the cursor to highlight $y1$. Then press [2nd] [F6] to display the Style menu. Choose item 8, Below, by pressing 8. (To shade above a line we would press 7 to select Above.) Then press [F2] 6 to see the graph of the inequality in the standard viewing window.

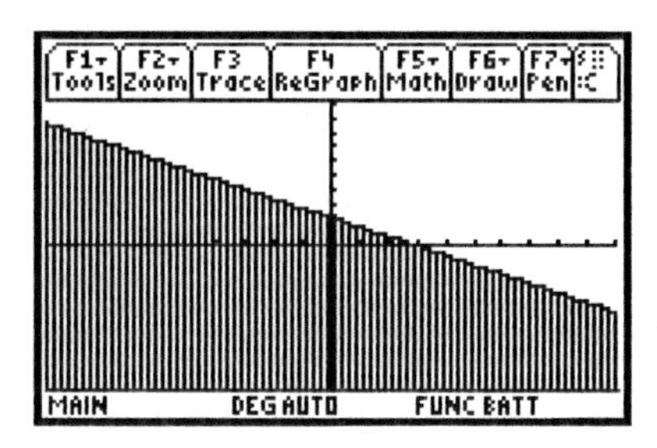

SYSTEMS OF LINEAR INEQUALITIES

We can graph systems of inequalities by shading the solution set of each inequality in the system with a different pattern. When the "Above" or "Below" style options are selected, the calculator rotates through four shading patterns. Vertical lines shade the first equation, horizontal lines the second, negatively sloping diagonal lines the third, and positively sloping diagonal lines the fourth. These patterns repeat if more than four equations are graphed.

Section 4.6, Example 3 Graph the solutions of the system

$$x - 2y < 0,$$
$$-2x + y > 2.$$

First we solve $x - 2y < 0$ for y, obtaining $y > x/2$. We graph the boundary line $y = x/2$ and, since the inequality can be written in the form $y > mx + b$, we select the "Above" style for this entry. (See Example 5 above for instructions on selecting styles.) Next we solve $-2x + y > 2$ for y, obtaining $y > 2x + 2$. We see that this inequality can be written in the form $y > mx + b$, so we graph the boundary line $y = 2x + 2$ and shade above it. Trial and error shows that $[-7, 5, -6, 6]$ is a good viewing window for this system of inequalities. The graph shows the solution set of each inequality in the system and the region where they overlap. The region of overlap is the solution set of the system of inequalities. Keep in mind that neither boundary line is part of the solution set.

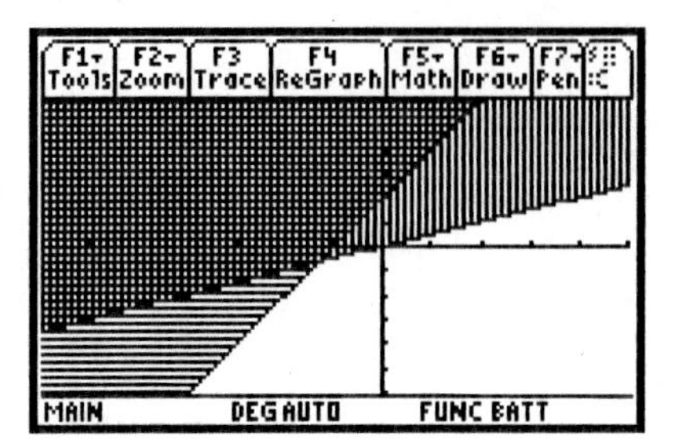

Chapter 5
Polynomials

EVALUATING POLYNOMIALS

Section 5.2, Example 6 Evaluate $-x^2 + 3x + 9$ for $x = -2$.

We can use a graphing calculator to evaluate polynomials in several ways. We can enter the polynomial on the home screen, substituting a value for the variable, as described on page 38 of this manual. We can also use a table set in Ask mode as described on page 10 of this manual.

In addition to the methods mentioned above, we can use a graph to evaluate a polynomial. First we graph $y_1 = -x^2 + 3x + 9$. The window $[-10, 10, -10, 15]$ is a good choice for this graph. Now we can use either the Value option from the Math menu on the Graph screen or the Trace feature to evaluate the polynomial. To use the Value option to find the value of y_1 when $x = -2$, on the Graph screen first press F5 1 to select Value. Then press (−) 2 ENTER. To use Trace to evaluate the polynomial after it is graphed, from the Graph screen press F3 to select Trace. Then use the up or down arrow key to place the cursor on the curve. Now enter the desired value for x, -2. In each case we see xc:-2, yc:-1 at the bottom of the screen. This indicates that the value of $-x^2 + 3x + 9$ is -1 when $x = -2$.

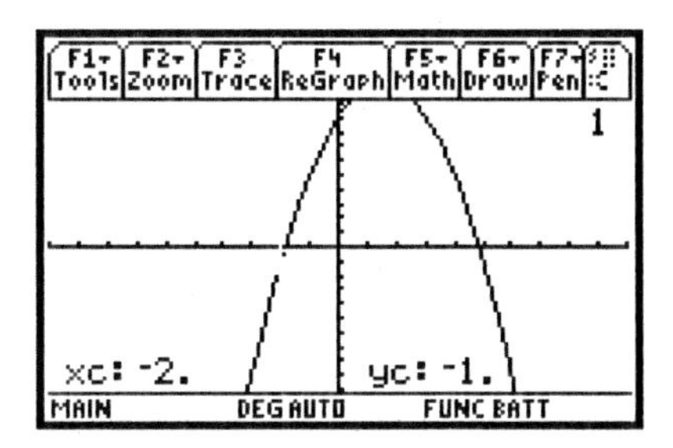

When using the Value feature, note that the x-value entered must be in the viewing window. That is, x must be a number between xmin and xmax.

CHECKING OPERATIONS ON POLYNOMIALS

A graphing calculator can be used to check operations on polynomials.

Section 5.3, pages 320 and 321 Several methods for checking whether two algebraic expressions are equivalent are discussed on pages 320 and 321 of the text. We will illustrate these methods by checking the addition $(-3x^3 + 2x - 4) + (4x^3 + 3x^2 + 2) = x^3 + 3x^2 + 2x - 2$.

First we will check this result by comparing the graphs of $y1 = (-3x^3 + 2x - 4) + (4x^3 + 3x^2 + 2)$ and $y2 = x^3 + 3x^2 + 2x - 2$. This is most easily done when different graph styles are used for the graphs. The **path graph style** can be used, along with the line style, to determine whether graphs coincide. First, on the equation-editor screen, enter $y1 = (-3x^3 + 2x - 4) + (4x^3 + 3x^2 + 2)$ and $y2 = x^3 + 3x^2 + 2x - 2$. We will select the path style from the style menu for $y2$. To do this, highlight the expression for $y2$ and then press 2nd F6 6 ENTER or press 2nd F6 ▽ ▽ ▽ ▽ ▽ ENTER.

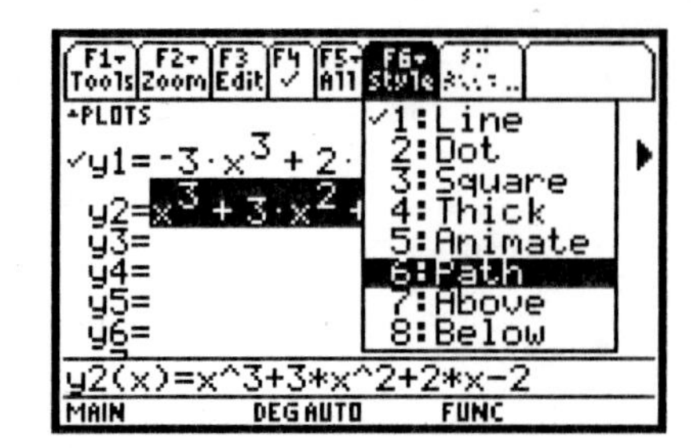

The calculator will graph $y1$ first as a solid line. Then $y2$ will be graphed as the circular cursor traces the leading edge of the graph, allowing us to determine visually whether the graphs coincide. In this case, the graphs appear to coincide, so the factorization is probably correct.

We can also check the addition by **subtracting** the result from the original sum. With $y1$ and $y2$ entered as described above, position the cursor beside "$y3 =$" and enter $y_3 = y1 - y2$ by pressing Y 1 − Y 2 ENTER. If the subtraction is correct, $y1 = y2$, so $y3 = 0$ and the graph of $y3$ will be $y = 0$, or the x-axis.

Since we are interested only in values of $y3$, we will deselect $y1$ and $y2$. To do this, first highlight the expression entered for $y1$ and press F4. Repeat this for $y2$. Note that there are no longer check marks to the left of these equations, indicating that the equations have been deselected. The graph of an equation that has been deselected will not appear when ◇ GRAPH is pressed. A deselected equation can be selected again by highlighting the equation and pressing F4. Note that a check mark once again appears to the left of the equation.

Select the path graph style for $y3$ as described above.

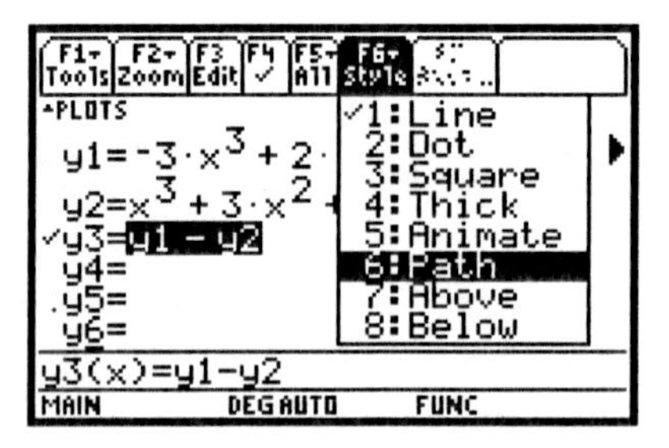

Now press ◇ GRAPH and determine if the graph of $y3$ is traced over the x-axis. Since it is, the sum is correct.

We can also use a table of values to **compare values** of $y1$ and $y2$. If the expressions for $y1$ and $y2$ are the same for each given x-value, the result checks. If you deselected $y1$ and $y2$ to check the sum using subtraction above, select them again as described above now. Then look at a table set in Auto mode. (See page 45.) Since the values of $y1$ and $y2$ are the same for each given x-value, the result checks. Scrolling through the table to look at additional values makes this conclusion more certain.

x	y1	y2
-3.	-8.	-8.
-2.	-2.	-2.
-1.	-2.	-2.
0.	-2.	-2.
1.	4.	4.

x= -3.
MAIN DEG AUTO FUNC

We can also check the sum using a **horizontal split screen**, or **top-bottom split screen**. We will choose to display the graph in the top half of the screen and a table of values in the bottom half.

First enter $y1$ and $y2$ as described above. Then set up a table in Auto mode. We will use tblStart $= -3$ and Δtbl $= 1$. To select the top-bottom split-screen option, first press MODE F2 to access the second page of the Mode screen. Then press ▷ 2 to select a screen split into a top and a bottom portion. Next press ▽ ▷ 4 ▽ ▷ 5 ENTER to select the graph to be displayed in the top portion of the screen and a table of values in the bottom portion. It might be necessary to press ◇ TABLE to see the table. We show the equations graphed in the standard window.

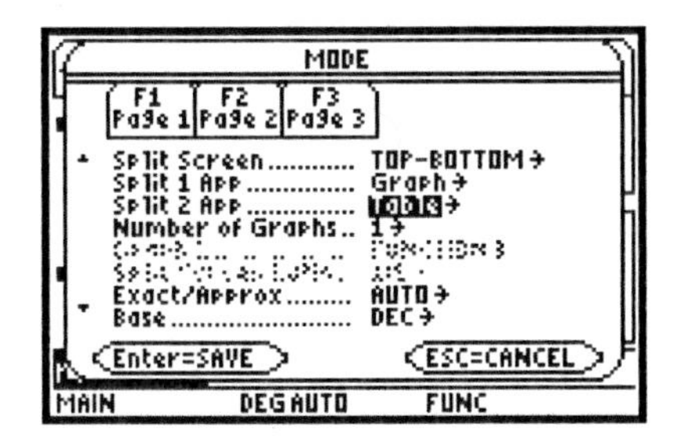

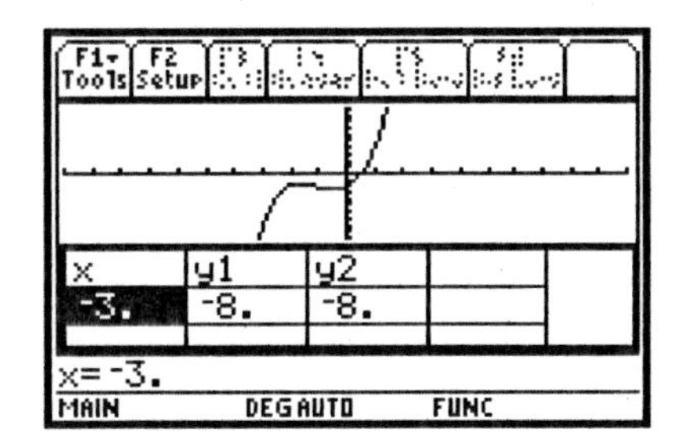

We can also use a **vertical split screen**, of **left-right split screen** to check the addition. First enter $y1$, $y2$, and $y3$ as described above and deselect $y1$ and $y2$. Then press MODE F2 to display the second Mode screen. Press ▷ to select left-right. Then select Graph for Split 1 App and Table for Split 2 App. Now press ENTER to see the graph of $y3$ on the left side of the screen and a table of values for $y3$ on the right side. As we did above, we use show the graph in the standard window and use a table set in Auto mode with tblStart $= -3$ and Δtbl $= 1$. Since the graph appears to be $y = 0$, or the x-axis, and all of the $y3$-values in the table are 0, we confirm that the result is correct.

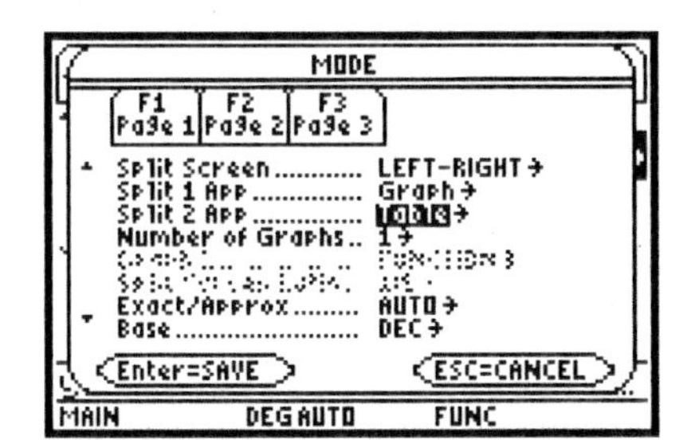

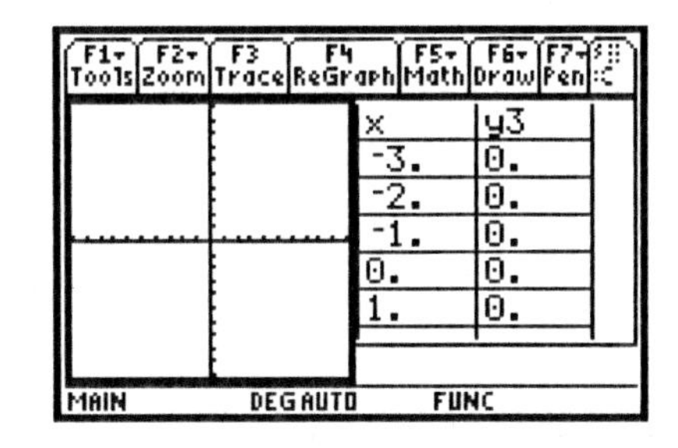

In order to return to a full screen, return to the second Mode screen and select Full for the Split Screen option.

EVALUATING POLYNOMIALS IN SEVERAL VARIABLES

Section 5.6, Example 1 Evaluate the polynomial $4 + 3x + xy^2 + 8x^3y^3$ for $x = -2$ and $y = 5$.

To evaluate a polynomial in two or more variables, substitute numbers for the variables. This can be done by substituting values for x and y directly or by storing the values of the variables in the calculator. The procedure for direct substitution is described on page 38 of this manual.

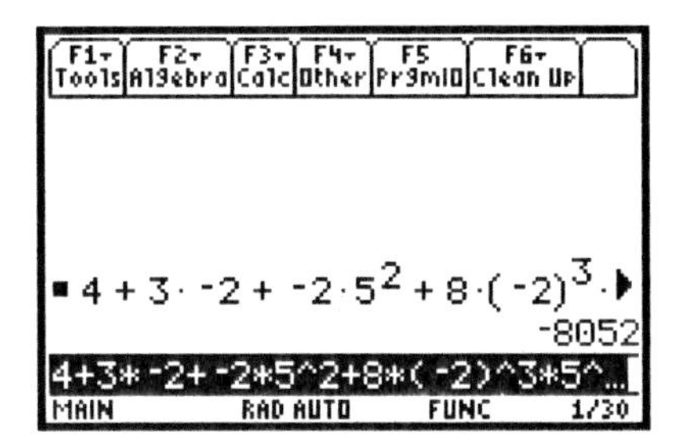

To evaluate the polynomial by first storing -2 as x and 5 as y, we proceed as follows. To store -2 as x, press (−) 2 STO▷ X

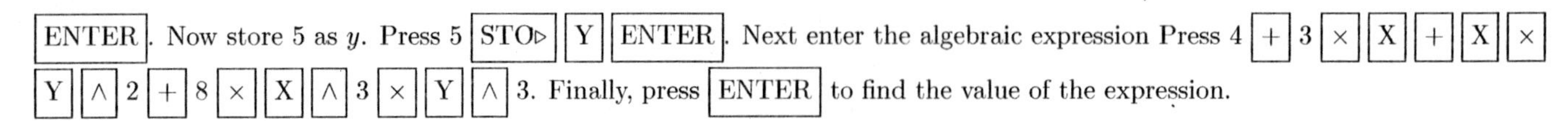

ENTER . Now store 5 as y. Press 5 STO▷ Y ENTER . Next enter the algebraic expression Press 4 + 3 × X + X × Y ∧ 2 + 8 × X ∧ 3 × Y ∧ 3. Finally, press ENTER to find the value of the expression.

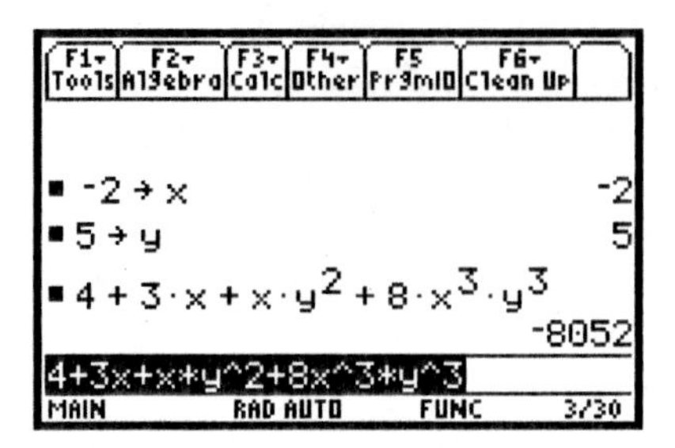

SCIENTIFIC NOTATION

To enter a number in scientific notation, first type the decimal portion of the number; then press the EE key in the left column of the keypad; finally type the exponent, which can be at most three digits. For example, to enter 1.789×10^{-11} in scientific notation, press 1 . 7 8 9 EE (−) 1 1 ENTER . To enter 6.084×10^{23} in scientific notation, press 6 . 0 8 4 EE 2 3 ENTER . The decimal portion of each number appears before a small E while the exponent follows the E.

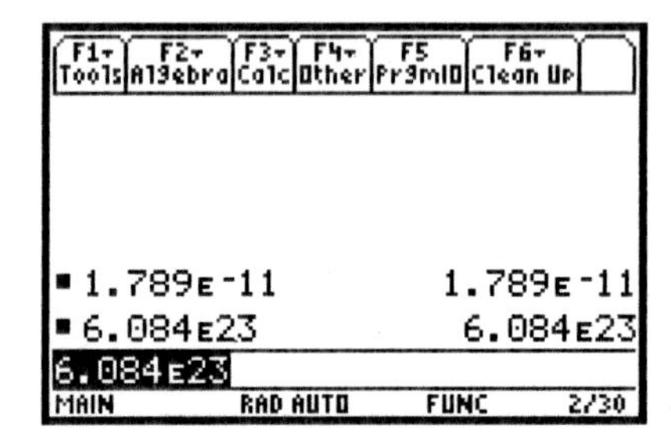

The graphing calculator can be used to perform computations in scientific notation.

Section 5.8, Example 9(b) Use a graphing calculator to check the computation $(7.2 \times 10^{-7}) \div (8.0 \times 10^{6}) = 9.0 \times 10^{-14}$.

We enter the computation in scientific notation on the entry line of the home screen. Press 7 . 2 EE (−) 7 ÷ 8 EE 6 ENTER . We have 9×10^{-14}, which checks.

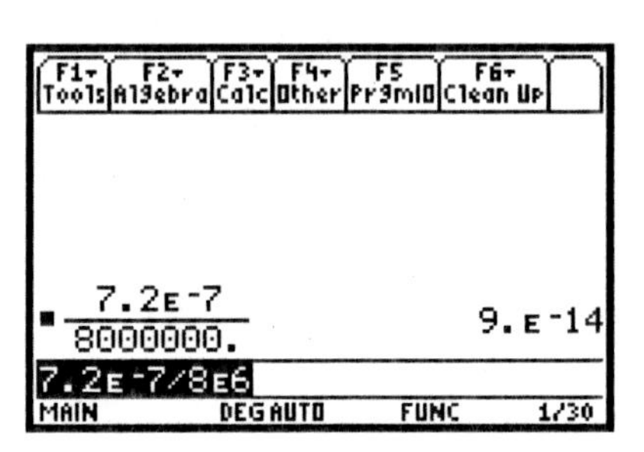

Chapter 6
Polynomials and Factoring

CHECKING FACTORIZATIONS OF POLYNOMIALS

Tables of values are used to check some of the factorizations in **Sections 6.1, 6.2, and 6.3**. The procedure for setting up tables like these is described on page 45 of this manual.

A graph drawn using different GraphStyles is used to check **Example 4** in **Section 6.2**. The procedure for doing this is found on page 55 of this manual.

A horizontal split screen, or top-bottom split screen, showing a graph and a table is used to check **Example 7** in **Section 6.4**. The procedure for setting up a horizontal split screen is described on pages 56 and 57 of this manual.

SOLVING QUADRATIC EQUATIONS

In **Section 6.6, page 420**, the ZERO option from the CALC menu is used to approximate the solutions of a quadratic equation. The procedure for using this method to solve an equation is found on page 50 of this manual.

Chapter 7
Rational Expressions and Equations

GRAPHING DATA

We use the Data/Matrix editor to enter data which can then be graphed.

Section 7.8, Example 6(a) The f-stop is a setting on a camera that indicates how much light will reach the exposed film. An f-stop is related to the size of the opening of the lens, or aperture. Some f-stops and apertures for a 50 mm lens are shown in the following table. Graph the data and determine whether the data indicate direct variation or inverse variation.

f-stop	Size of aperture, in mm
1.4	35.7
4	12.5
8	6.25
11	4.545

We will enter the data as ordered pairs in the Data/Matrix editor. Press APPS 6 3 to display the new data variable screen in the Data/Matrix editor. We must now enter a data variable name in the Variable box on this screen. The name can contain from 1 to 8 characters and cannot start with a numeral. Some names are preassigned to other uses on the TI-89. If you try to use one of these, you will get an error message. Press ▽ ▽ to move the cursor to the Variable box. We will name our data variable "fstop." To enter this name, first lock the alphabetic keys on by pressing 2nd a-lock. Then press F S T O P. Note that F, S, O, and P are the purple alphabetic operations associated with the |, 3, (−), and STO▷ keys, respectively. The T key appears on the keypad to the left of the ∧ key.

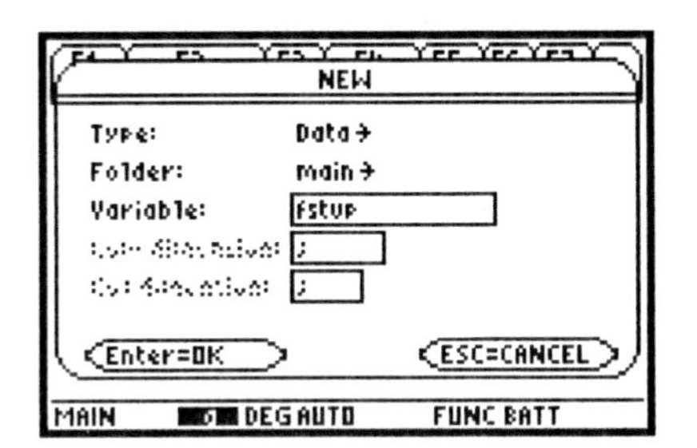

After typing the name of the data variable, unlock the alphabetic keys by pressing the purple alpha key. Now press ENTER ENTER to go to the data-entry screen. Assuming the data variable name "fstop" has not previously been used in your calculator, this screen will contain empty data lists with row 1, column 1 highlighted. If entries have previously been made in a data variable named "fstop," they can be cleared by pressing F1 8 ENTER.

We will enter the first coordinates in column c1 and the second coordinates in c2. To enter 1.4, press 1 . 4 ENTER. Continue typing the first coordinates 4, 8, and 11, each followed by ENTER. The entries can be followed by ▽ rather than ENTER if desired. Press ▷ △ △ △ △ to move to the top of column c2. Type the second coordinates 35.7, 12.5, 6.25, and 4.545 in succession, each followed by ENTER or ▽. Note that the coordinates of each point must be in the same position in both lists.

F1 Tools | F2 Plot Setup | F3 Cell | F4 Header | F5 Calc | F6 Util | F7 Stat

DATA

	c1	c2	c3
1	1.4	35.7	
2	4	12.5	
3	8	6.25	
4	11	4.545	

r1c3=

MAIN DEG AUTO FUNC BATT

Now press ◇ Y = to go to the equation-editor screen and clear any equations that are currently entered. (See page 43 of this manual.) If you wish, instead of clearing an equation, you an deselect it. (See page 56 of this manual.) In order to graph the data points we will turn on and define a plot. To do this, from the STAT VARS screen first press ENTER F2 to go to the Plot Setup screen. We will use Plot 1, which is highlighted. If any plot settings are currently entered beside "Plot 1,"" clear them by pressing F3. Clear settings shown beside any other plots as well by using ▽ to highlight each plot in turn and then pressing F3.

Now we define Plot 1. Use △ to highlight Plot 1 if necessary. Then press F1 to display the Plot Definition screen. The item on the first line, Plot Type, is highlighted. We will choose a scatter diagram, denoted by Scatter, by pressing ▷ 1. Now press ▽ to go to the next line, Mark. Here we select the type of mark or symbol that will be used to plot the points. We select a box by pressing ▷ 1. Now we must tell the calculator which columns of the data variable to use for the x- and y-coordinates of the points to be plotted. Press ▽ to move the cursor to the "x" line and enter c1 as the source of the x-coordinates by pressing alpha C 1. (C is the purple alphabetic operation associated with the) key.) Press ▽ alpha C 2 to go the the "y" line and enter c2 as the source of the y-coordinates.

To select the dimensions of the viewing window notice that the f-stops in the table range from 1.4 to 11 and the aperture size ranges from 4.545 to 35.7. We want to select dimensions that will include all of these values. One good choice is [0, 12, 0, 50], yscl = 10. Enter these dimensions in the WINDOW screen. Then press ◇ GRAPH to graph the data. From the graph we see that the data indicate inverse variation.

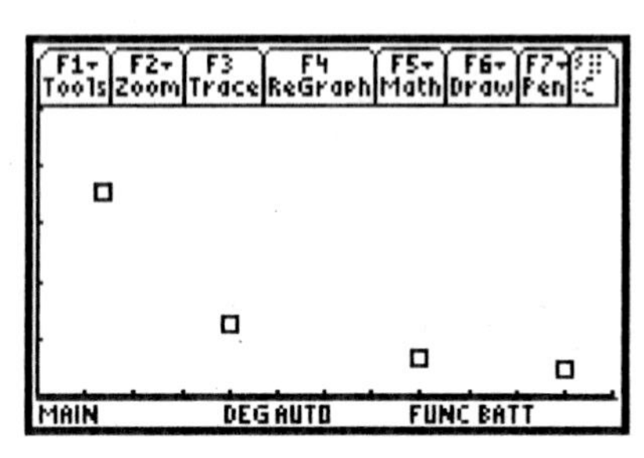

Chapter 8
Radical Expressions and Equations

FINDING SQUARE ROOTS

Section 8.1, Example 5 Use a calculator to approximate $\sqrt{10}$ to three decimal places.

To find a decimal approximation for $\sqrt{10}$, first select Approximate mode. Then enter $\sqrt{10}$ by pressing [2nd] [$\sqrt{\ }$] 1 0 [)] [ENTER]. ($\sqrt{\ }$ is the second operation associated with the [×] multiplication key.) Note that the calculator supplies a left parenthesis along with the radical symbol. On the TI-89 we must close the parentheses by adding a right parenthesis after entering the radicand. We see that $\sqrt{10} \approx 3.162$.

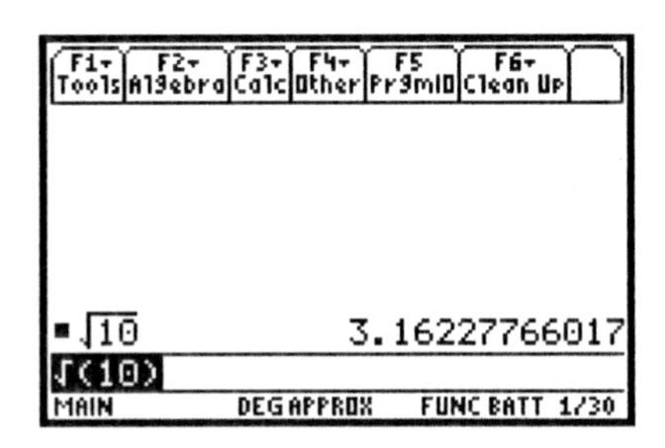

Chapter 9
Quadratic Equations

EVALUATING A FUNCTION

Using function notation to evaluate a function is discussed in the text in **Section 9.7** on **page 612**. For example, to find $f(-1.5)$ when $f(x) = -2x^2 + 3x - 1.2$, first press [◇] [Y =] to go to the equation-editor screen. Then mentally replace $f(x)$ with $y1$ and enter $y1 = -2x^2 + 3x - 1.2$. Now, to find $f(-1.5)$, or $y1(-1.5)$, first press [2nd] [QUIT] to go to the home screen. Then enter "$y1(-1.5)$" on the entry line by pressing [Y] 1 [(] [(−)] 1 [.] 5 [)] [ENTER]. We see that $y1(-1.5) = -10.2$, or $f(-1.5) = -10.2$.

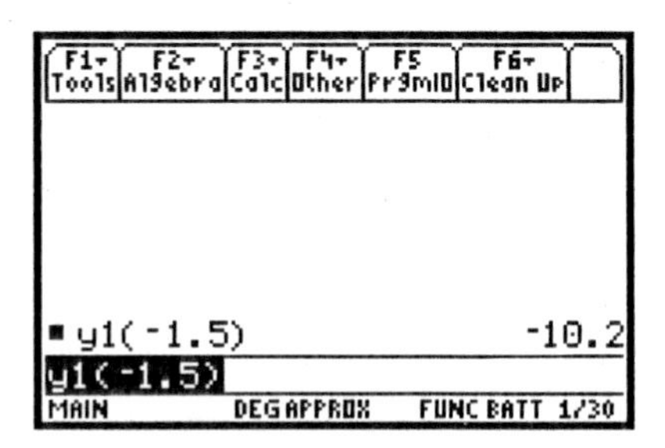

Index
TI-83, TI-83 Plus, and TI-84 Plus Graphics Calculators

Index
TI-89 Graphics Calculator

TI-89 Index